MEMS Barometers Toward Vertical Position Detection

Background Theory, System Prototyping, and Measurement Analysis

Synthesis Lectures on Mechanical Engineering

Synthesis Lectures on Mechanical Engineering series publishes 60–150 page publications pertaining to this diverse discipline of mechanical engineering. The series presents Lectures written for an audience of researchers, industry engineers, undergraduate and graduate students.

Additional Synthesis series will be developed covering key areas within mechanical engineering.

© Springer Nature Switzerland AG 2022

Reprint of original edition © Morgan & Claypool 2017

MEMS Barometers Toward Vertical Position Detection:
Background Theory, System Prototyping, and Measurement Analysis
Dimosthenis E. Bolanakis

ISBN: 978-3-031-79572-5 paperback
ISBN: 978-3-031-79573-2 ebook

DOI 10.1007/ 978-3-031-79573-2

A Publication in the Springer series
SYNTHESIS LECTURES ON MECHANICAL ENGINEERING

Lecture #3
Series ISSN
ISSN pending.

MEMS Barometers Toward Vertical Position Detection

Background Theory, System Prototyping,

and Measurement Analysis

Dimosthenis E. Bolanakis

Department of Air Force Science (Informatics and Computers section)

SYNTHESIS LECTURES ON MECHANICAL ENGINEERING #3

ABSTRACT

Micro-Electro-Mechanical-Systems (MEMS) sensors constitute perhaps the most exciting technology of our age. The present effort incorporates all the information needed by scientists and engineers who work on research projects and/or product systems, which apply to air pressure acquisition and to its rearrangement into altitude data. Some of the potential implementations of this method (regularly referred to as barometric altimetry) include, but are not limited to, *Position Location Application, Navigation Systems, Clinical Monitoring Applications*, and *Aircraft Instrumentation.*

This book holds the key to such applications, providing readers with the theoretical basis as well as the practical perspective of the subject matter. At first, the reader is introduced to the background theory, methods, and applications of barometric altimetry. Thereafter, the book incorporates the development of wireless barometers and a (real-time monitoring) wireless sensor network system for scheduling low-cost experimental observations. Finally, a deepened understanding to the analysis procedure of pressure measurements (using Matlab script code) is performed. Some accompanying material can be found at `http://bit.ly/mems-files`.

KEYWORDS

MEMS, barometers, altimetry, barometric pressure sensors, system prototyping, measurement analysis

I've never seen any of the 7 wonders of the world and from any available list. I believe there are more though… I see perfection and plenty in number 10 everyday…

To my wife, son, and daughter

Contents

List of Figures

List of Formulas

List of Tables

Preface

Micro-Electro-Mechanical-Systems (MEMS) sensors constitute perhaps the most exciting technology of our age. As the acronym implies, MEMS devices integrate a mechanical along with an electrical feature embedded into the same structure die. The physical change in the mechanical function is converted into electrical signal. For instance, a deflection to the sensing membrane of a piezoresistive pressure sensor results in the change of the measured (output) resistance. Today's commercial MEMS sensors encompass both sensing and *integrated circuit (IC)* die within the same package. The IC die is addressed to provide further processing to the sensing element (e.g., interpreting the electrical signal to a serial computer interface such as the I2C). Accordingly, the modern MEMS sensor technology is delivered in the form of a *system in package (SiP)* module, which could be easily interfaced by an embedded computer system, such as a *microcontroller unit (MCU)*.

The unique advantages of MEMS sensors technology are in agreement with a miniaturized-size device of low-power consumption, low cost and particularly high performance. Due to these beneficial features, different types of MEMS sensors (e.g. accelerometers, gyroscopes, etc.) can nowadays be found in an abundance of electronics devices, including the most popular ones like smartphones and tablet PCs. Accordingly, there is particular research based on the utilization MEMS sensors found in consumer devices. A prime example of such a case is the exploitation of MEMS barometric pressure sensors toward the determination of vertical displacements in position location systems. However, there are particular limitations in utilization of the existing mobile technology.

The present effort incorporates all the information needed by scientists and engineers who work on research projects and/or product developments of systems, which apply to air pressure acquisition and to its rearrangement into altitude data. This method is also known as *barometric altimetry*. Some of the potential implementations of barometric altimetry are as follows:

a) Position location application (e.g., the floor detection inside large buildings, such as museums, is addressed for delivering information to the user's mobile phone about the exhibits available on the identified floor);

b) Navigation systems (e.g., sensor fusion techniques on barometric and geometric altitude reading are intended for improving GPS navigation);

c) Clinical monitoring applications (e.g., vertical displacements identification is related to the fall of patient);

d) Aircraft instrumentation (e.g., vertical displacements are interpreted into the aircraft's pitch and roll positional angles).

This book holds the key to such applications, providing readers with the theoretical basis as well as the practical perspective of the subject matter. In detail, the first section introduces readers to the background theory, methods, and applications of barometric altimetry. The second section contributes to the development of wireless barometers and a (real-time monitoring) wireless sensor network system. The concluding section provides a deepened understanding to the analysis procedure of pressure measurements, using Matlab script code.

Barometric altimetry features two main problems that severely affect accuracy of measurement. The former is induced by the ambient air pressure changes, as time elapses. The latter is in agreement with the manufacturing dissimilarities of identical MEMS sensor devices, which generate a slightly different measurement upon the sensing of the exact same pressure value. The book deals with those issues and addresses innovative techniques for scheduling low-cost experimental observations of the system performance, over the conventional and exceptionally expensive apparatus required for conducting accurate experiments. The book contributes to the methods and techniques for sustaining long-term measurement stability.

Such strategies could also be addressed for laboratory training on MEMS pressure sensors and barometric altimetry in the graduate level, while they could be easily adopted by the departments of Electrical and Electronic Engineering and Physics. The present project aims in extending prevailing possibilities for research and laboratory training on MEMS barometers, and provides readers with an integrated theoretical and practical learning framework.

ORGANIZATION OF THE BOOK

This book consists of four chapters and one appendix, developed to support the theoretical and practical part of MEMS pressure sensors and barometric altimetry applications. The following table summarizes contents of the book.

Chapter 1: Principles, Background Theory, and Applications

- Introduction to MEMS sensors

- MEMS-based product development from scratch

- MEMS pressure sensors and characterization techniques

- Atmospheric pressure measurements and barometric altimetry

- References

Chapter 2: System Prototyping

- Single-ended altimetry measurement setup

- Differential altimetry measurement setup

- Software interfacing

- Arrangement of low-cost experimental observations

- References

Chapter 3: Effects on Measurement Accuracy

- Gaussian white noise signal in barometric sensor output signal

- Temperature influences on barometric altitude measurements

- Influences of the existing pressure level on accuracy

- References

Chapter 4: Absolute Height Acquisition & Measurement Analysis

- Measurement analysis in differential altimetry

- Measurement analysis in single-ended altimetry

- Statistical significance of the estimator

- References

Appendix A

- Measurement analysis of temperature influence (Matlab code)

- Measurement analysis of pressure influence (Matlab code)

- Statistical analysis of height measurements (Matlab code)

- References

Chapter 1 introduces readers to MEMS & sensors technology and explores the background theory, methods, and applications of barometric altimetry toward vertical (absolute) height determination from ambient air pressure measurements.

Chapter 2 presents the system design and development of wireless barometers and a (real-time monitoring) wireless sensor network employing MEMS sensors.

Chapters 3 and 4 provide deepened understanding to the pressure measurement and statistical analysis of the estimator (analysis is performed in Matlab).

The appendix provides additional information to the Matlab Script code addressed by the book.

Dimosthenis E. Bolanakis
May 2017

CHAPTER 1

Principles, Background Theory, and Applications

This chapter introduces readers to the MEMS sensors technology, and also explores the background theory and applications of barometric altimetry.

1.1 INTRODUCTION TO MEMS SENSORS

Micro-Electro-Mechanical-Systems (MEMS), as defined by Wikipedia [1], is the technology of microscopic devices featuring moving parts. The keywords in this definition are "microscopic devices" and "moving parts." The former denote a technology involving a microscope or, more precisely, a stereo microscope (also known as *stereoscope*[1]). The latter refers to a mechanical element integrated, along with an electrical component, into the MEMS device. The physical change in the mechanical function is converted into an electrical signal. Accordingly, the MEMS acronym implies the integration (and interdependence) of a mechanical and electrical feature embedded into a miniaturized structure *die*; that is, a small block of material cut ("diced") out from a wafer (Figure 1.1).

The unique advantages of MEMS sensors result from their fabrication techniques. Many of them are the same as in the *integrated circuits* (ICs) fabrication, but they are also combined with micromachining processes (e.g., etching parts away of the silicon wafer) in order to realize the miniaturized mechanical elements. The micromachined fabrication offers increased performance and low-power consumption over the conventional macro-scale techniques [2], while the regular IC fabrication provides a low production cost. Subsequently, the beneficial features of a miniaturized-size device of low-power consumption, low cost, and particularly high performance, have brought a new direction in industry over the last decade [3].

Before exploring the great challenges and opportunities present in MEMS-based applications, it would be wise to provide a brief overview to the typical procedure followed toward the product design and development of this microscopic sensor technology, when starting from scratch. The overall production processes presented hereafter reveal the interdisciplinary "nature" of this MEMS sensors technology. Some of the involved technical areas include, but are not limited to, material science, silicon design, packaging, and instrumentation and measurement.

[1]Comparatively to the light microscope, a stereoscope provides the further ability of perceiving depth, in terms of a three-dimensional visualization.

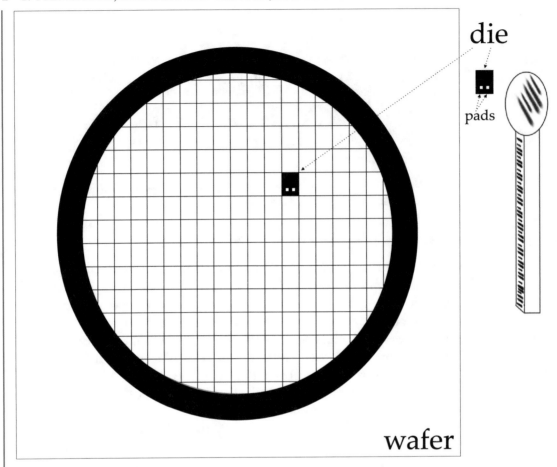

Figure 1.1: MEMS die cut out from a wafer (smaller than the size of a matchstick head).

1.2   MEMS-BASED PRODUCT DEVELOPMENT FROM SCRATCH

MEMS physical design requires a *computer aided design* (CAD) tool (e.g., CoventorWare [4]), which enables simulation as well as layout modeling of the device. Having verified all aspects of the prototype MEMS device, the designers forward the layout files to fabrication. There are particular foundries manufacturing silicon wafers using the existing IC fabrication techniques, which are also supplemented with micromachining processes (such as, X-FAB [5]). The final wafer is delivered to the customer and a series of verification procedures are performed in the appropriate laboratory so as to assess whether the prototype MEMS sensor meets the initial design requirements and specifications. Those tests are typically conducted within a clean room of low-level environmental pollutants (e.g., dust).

Figure 1.2 presents the laboratory equipment required for a wafer-level testing of the prototype MEMS sensor device. This setup is used for measuring the electrical signal of the sensor (regularly resistance or capacitance) at room conditions. The probe system supports a micrometer-scale resolution of movement of the wafer chuck and probes. Beneath the stereoscope, the probes are attached to the sensor pads and a high-precision measurement is acquired by an external multimeter. Not every sensor in the wafer is functional and therefore, this measurement is needed in order to select the proper sensor before proceeding to the die and wire-bonding processes.

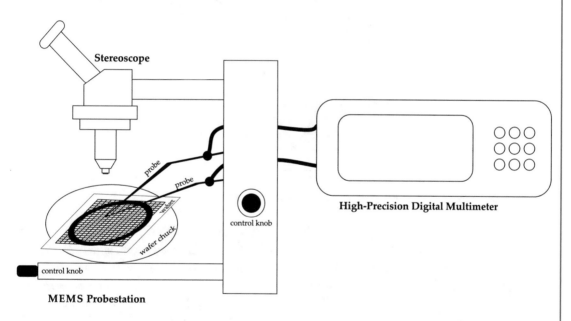

Figure 1.2: Wafer-level MEMS testing.

Die bonding is the process of picking a sensor die from the wafer and permanently attaching it to a target substrate, such as *printed circuit board* (PCB), MEMS package, etc. This process is performed though a (vacuum) pick and place machine. Alignment of the sensor to the substrate is assured though a stereoscopic observation of the procedure. For a research laboratory, the bonding attachment technique could rely on a heat-curing adhesive. In detail, a small drop of adhesive is dispensed on the substrate and then, the die is placed on top of it with the use of the pick and place machine. Finally, the curing process is accomplished through the heating of substrate at an elevated temperature (specified by the adhesive's technical datasheet), so as to harden the material and permanently mount the sensor die on the PCB or package.

Wire bonding is the method of interconnecting the MEMS sensor to the PCB or package, and it is typically performed using either (a) gold or (b) aluminum wire of diameter equal to a few tens of micrometers. This method is also supported by a stereoscopic observation and it is performed via a special machine called wire bonder. The wire bonder is usually compared to a sewing machine, as the wires are disposed to the die/substrate pads though a needle-like tool called *capillary*. The machine applies ultrasonic energy and bonding force in order to accomplish the wire connection. Aluminum wire connections are of tail-to-tail bonds, while gold wire connections are of ball-to-tail bonds (or ball-to-ball bonds for more solid connection).

As soon as the wire bonding finishes, a glob top sealing compound can be used for the protection of bonds against mechanical strains, moisture, etc. A thermally curing glob top could be dispensed and fully cover the wire bonds (and ICs in some cases), and then heat-cured at a particular temperature. Figure 1.3 presents a gold wire (ball-to-tail) interconnection between a sensor die and a substrate, which is fully covered by glob top.

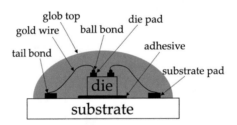

Figure 1.3: Wire bonding and glob top sealing compound.

Today's MEMS sensors encompass the sensor die along with the IC die. The latter is addressed for the implementation of further processing to the MEMS electrical signal. Particular processing examples are in line with (a) temperature compensation of the pressure output signal, (b) interpretation of the electrical signal to a serial computer interface, such as *serial peripheral interface* (SPI) or *inter–integrated circuit* (I^2C), etc. Figure 1.4 depicts the top, bottom, and inside view of a possible package incorporating a *barometric pressure sensor* (BPS) along with the IC. A hole on the top of the package (Figure 1.4a) allows the air molecules to reach the membrane of BPS, while the four pads at the bottom of the package (Figure 1.4b) render feasible the interconnection of BPS with an embedded system (such as, a microcontroller) via I^2C interface. There are particular corporations that work on packaging so as to integrate the MEMS device into a *system in package* (SiP) module, such as the Advanced Semiconductor Engineering, Inc. [6], specializing in *wafer–level packaging* (WLP) solutions. The latter technology refers to the slicing of a wafer into individual dies and thereafter packaging them using the aforementioned methods.

Once reaching this point of the MEMS-based product development procedure, the subsequent step is in agreement with particular experiments applying to sensors characterization. Those experiments require a specific apparatus, peculiar to the type of MEMS sensor being characterized (e.g., pressure sensor, accelerometer, etc.). Particular aspects of the sensor are de-

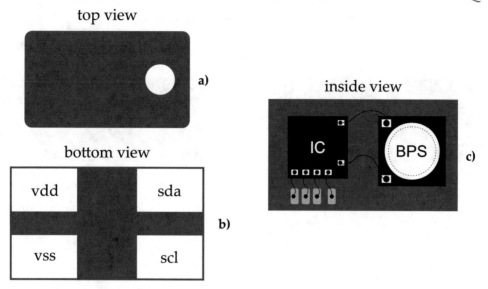

Figure 1.4: Top, bottom, and inside view of a possible package of barometric pressure sensor (BPS).

termined (such as, accuracy, hysteresis, repeatability, etc.) in relation to the sensor's output signal and input stimulus [7]. Hereafter, we explore the fundamental types of MEMS pressure sensors as well as the experimentation setup required for MEMS (barometric) pressure sensors characterization.

1.3   MEMS PRESSURE SENSORS AND CHARACTERIZATION TECHNIQUES

A preliminary categorization of pressure sensors is relevant to their sensing methods; that is, (a) piezoresistive and (b) capacitive sensors. A secondary strategy of classification is in agreement with the reference pressure, relative to which the value of sensing pressure is determined, that is (a) absolute, (b) gauge, and (c) differential pressure sensors. Figure 1.5 depicts the typical schematic cross section of a pressure sensor. In detail, a cavity is etched away of the silicon wafer in order to create a thin deformable diagram, which deflects under external pressure. The reference pressure can be either a vacuum sealed (i.e., zero pressure) or a backside port (created though a hall in the wafer). The former (Figure 1.5c,d) outlines an absolute pressure sensor. The latter (Figure 1.5a,b) forms either a gauge pressure sensor, in case the backside port is referenced to the atmosphere, or a differential pressure sensor, in case the measurement is determined by the difference of two pressures (i.e., the external and reference). Diaphragms of low/high thickness are proportional to the construction of low/high pressure sensor devices.

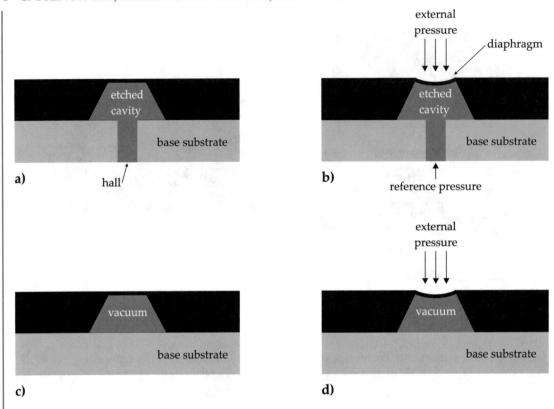

Figure 1.5: Typical schematic cross section of a MEMS pressure sensor.

In piezoresistive pressure sensors, the diaphragm is fabricated with four piezoresistive elements in a Wheatstone bridge. When the diaphragm deforms, it changes the resistance of these elements (i.e., increasing or decreasing resistance). In capacitive pressure sensors, one plate of a capacitor is attached to the diaphragm, while the other is fixed to the substrate. As diaphragm flexes in response to the external applied pressure, the distance in between plates changes and produces a variable capacitance. Performance specifications of a MEMS pressure sensor device are determined by a series of experiments using a specialized apparatus. Relevant definitions and characterization techniques are presented hereafter.

The most important feature of a sensor is *accuracy*, which is actually referred to device's *inaccuracy* [7], that is, the maximum deviation (i.e., measuring error) in sensor's output signal from its true input value (Figure 1.6a). A commonly accepted approach of defining accuracy in a MEMS pressure sensor is achieved by the sum of *nonlinearity*, *hysteresis*, and *nonrepeatability* errors at room temperature [8]. These error parameters are also known as *static error band*. *Nonlinearity* is the maximum deviation of a sensor's output from a reference straight line, normally defined by the *best-fit straight-line* (BFSL) method, when the output is measured with increasing pressure only (Figure 1.6b). Nonlinearity is regularly defined as ±x%FS, where *FS* is the *full scale (span)* pressure range. For example, if the sensor's operating pressure range is among 300–1,100 mbar (i.e., FS is equivalent to 800 mbar) and its nonlinearity is of ±0.2%, then the output varies as much as ±1.6 mbar of what the expected value should be. *Hysteresis* is the maximum deviation of a sensor's output, when the applied (input) pressure to the device is gradually increasing up to the top limit of FS, then decreasing back to the bottom limit (Figure 1.6c). Hysteresis is regularly referred to as *less than* x%FS (e.g., less than 0.02%FS). *Nonrepeatability* is the maximum deviation of a sensor's output, when the same input pressure is applied to the device twice, and from the same direction (Figure 1.6d). Nonrepeatability is regularly referred to as *within* x%FS (e.g., within 0.01%FS). Nonrepeatability and hysteresis are very major error parameters, as they reflect inherent quality of the sensor device, and it is important to be as low as possible [9].

Figure 1.7 presents the regular experimentation setup appropriate for MEMS absolute pressure sensors characterization. A source of clear dry nitrogen supplies, through a pressure regulator, the input of a pressure controller. Thereby the controller's output is able to provide a constant pressure level above the current atmospheric pressure. With the use of a vacuum pump connected to an auxiliary input of the controller, it is also possible to provide a constant pressure level below the existing barometric pressure. In consideration of the setup given in Figure 1.7, the controller keeps constant the inside pressure of an airtight enclosure, denoted "Controlled P". Inside the airtight enclosure are placed two *devices under test* (DUT); that is, two MEMS pressure sensors along with the necessary readout electronics (e.g., a microcontroller-based system) for sending the acquired pressure value to a host PC, denoted "Readout P". The setup employs an environmental chamber as well, which is addressed to control temperature (and/or humidity) of the airtight enclosure.

The environmental chamber is mainly needed because temperature has the largest effect on pressure measurement accuracy [10]. Therefore, particular techniques are required in order to minimize those effects and make available a SiP module that compensates for a specific temperature range. One possible solution is the evaluation of the sensor's raw data output (e.g., voltage) at several pressure levels (determined by the pressure controller), while repeating this procedure for different temperatures. Then, the calibration coefficients of a polynomial-based temperature compensation (of n*th* order for the raw measurement and m*th* order for the temperature) can be identified. Those coefficients are unique for each MEMS pressure sensor and they are typically stored into the IC embedded in the SiP module.

Two methods of providing the temperature errors over the compensated temperature range are the *total error band* (TEB) and *temperature coefficient* (TC). The former provides the error envelop that the MEMS pressure sensor is expected to operate within. The latter specifies a zero-point error at room temperature (e.g., 20°C) and an increasing (linear) error with the rises/falls of temperature. Figure 1.8 depicts TEB method (denoted within the grey area) as well as the TC method (given by the dotted line). Average zero-point temperature coefficient of this particular pressure sensor equals to 0.1%FS per 10 K.

Due to the need of providing temperature compensation to the MEMS (barometric) pressure sensors, the device embeds a temperature sensor within the SiP module [11, 12], as well. However, there are commercial sensors that provide designers with the option of reading raw data output and applying temperature compensation through particular math calculations (and

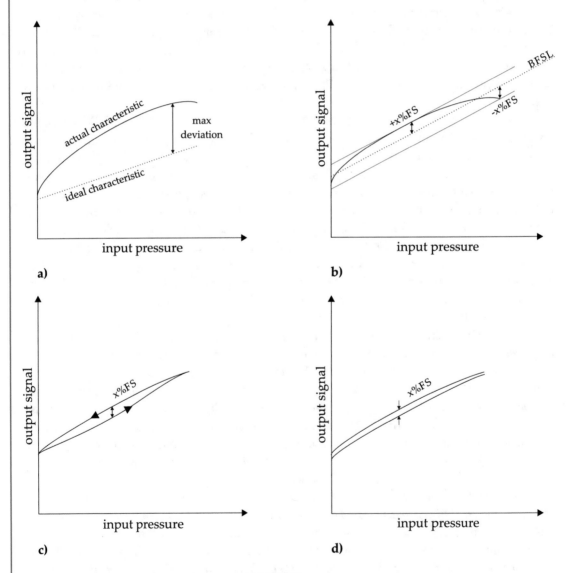

Figure 1.6: Performance specifications of a MEMS pressure sensor.

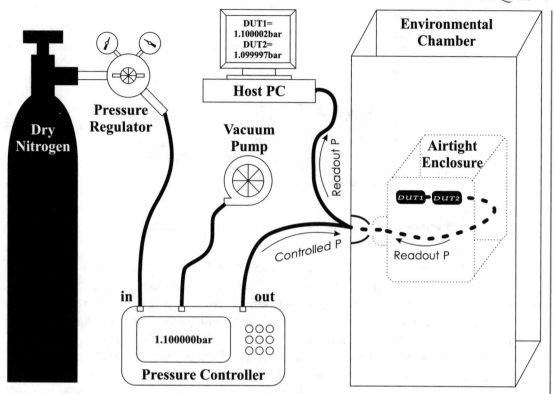

Figure 1.7: Experimentation setup for MEMS (barometric) pressure sensors characterization.

by using the unique coefficients stored into the sensor device) [11], while others provide the option of reading directly the compensated output [12]. The added benefit of the former approach against the latter is that it provides the possibility of arranging, if desired, extra temperature compensation to the raw pressure data. In addition, there are specific applications that require a rapid acquisition of pressure measurements. The possibility of reading raw data output and performing the actual result offline sustains this kind of application.

At this point of production, the MEMS sensor is ready to be employed in the embedded system; that is, for example, a microcontroller-based application for the acquisition of pressure data. At present, there are an abundance of cutting-edge applications applying to pressure sensors. For instance, Prisma Electronics SA [13] has developed a platform of wireless smart sensors addressed to monitor fuel consumption in vessel engines; Verimetra Inc. [14] has manufactured a scalpel able to sense the type of tissue being divided during a surgery (in consideration of the amount of force being applied to the blade); and several more examples.

For such applications, the arranging of research experiments is not easy as it requires the development of custom made apparatus. Alternative methods of scientific inquiries, in this field

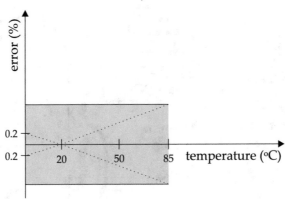

Figure 1.8: Temperature errors over the (temperature) compensated range.

of study, are possible with the organization of experiments that employ barometric pressure sensors and apply to altimetry. Recent applications in the literature exploit the MEMS barometric sensors found in smartphones, in order detect vertical displacements in position location systems. However, the utilization of the existing mobile technology features particular limitations. The proposed book provides readers with a particularly helpful guide toward the implementation of position location systems and applications in terms of vertical position/displacement detection, using MEMS barometric sensors attached to a *wireless sensor network* (WSN) system. Later in this chapter, readers are introduced to the background theory, methods, and applications of barometric altimetry.

1.4   ATMOSPHERIC PRESSURE MEASUREMENTS AND BAROMETRIC ALTIMETRY

Pressure (P) term defines the amount of *force* (F) per unit *area* (A); that is, $P = F/A$. Subsequently, the *atmospheric pressure* is a function of the gravitational force on air molecules, pushing them down on Earth. Because of the smaller amount of air molecules at higher elevation, air pressure is considered inversely proportional (though not linear) to the altitude. Particular features of the earth's atmosphere are as follows: (a) the atmosphere is not uniform, (b) it is in constant motion, and (c) wind blows from high to low pressure areas [15].

Atmospheric pressure measurements are regularly exploited by meteorologists, who typically normalize readings with the local sea pressure level in order to make predictions for the weather in the future. Another dominant application related to the air pressure acquisition is in line with its rearrangement into altitude data; a method regularly referred to as *barometric altimetry*. The electronic device addressed for the acquisition of atmospheric pressure is called a *barometer*, while the barometer calibrated to output altitude data is regularly referred to as a *barometric altimeter*.

Altimeters are typically employed in avionics (as well as consumer electronics, such as watches) in order provide information about how high above mean sea level is an aircraft (or mountain climber) located. This information is in reference to either the *international standard atmosphere* (ISA) of the *local sea-level pressure* (also known as QNH [16]). For the ISA model sea-level pressure and temperature are assumed to be fixed at the constant values $P_0 = 101325$ Pa and $T_0 = 15°C (= 288.15$ K), respectively. For an isothermal atmosphere the pressure decreases exponentially with height and for a given altitude z, the atmospheric pressure $P_{(z)}$ is determined by Formula 1.1.

$P_{(z)} = P_0 e^{-(Z/H)}$, *where* : $H = \dfrac{kT}{mg}$ (Pressure Scale Height)

P_0 (International Standard Atmosphere) = 101325 Pa

k (Boltzmann's Constant) = $1.38 \cdot 10^{-23}$ m$^2 \cdot$ kg $\cdot$ sec$^{-2} \cdot$K^{-1}

T (Temperature expressed in K)

m (Average mass of atoms) = $4.76 \cdot 10^{-26}$ kg

   (22% of O_2 and 78% of N_2) $\Rightarrow (0.22 \cdot 2 \cdot 2.67 \cdot 10^{-26}$ Kg $+ 0.78 \cdot 2 \cdot 2.3 \cdot 10^{-26}$ Kg)

g (Acceleration of gravity) = 9.81 m/sec^2

Formula 1.1: Atmospheric pressure of a given altitude.

The *scale height* (H) in this formula is unaffected for a particular temperature. In order to rearrange the atmospheric pressure measurement into altitude data, we solve Formula 1.1 for altitude z (and the outcome occurs in meters).

Formula 1.2 is also known as the *hypsometric equation*. In order to determine altitude from a particular measurement of atmospheric pressure via hypsometric equation, the additional acquisition of temperature is needed. The hypsometric equation works for an isothermal atmosphere and, therefore, an estimation of the average temperature between the values of T_0 and $T_{(z)}$ should be performed. A more empirical equation, independent of temperature measure-

$\ln (P_{(z)}) = \ln (P_0 e^{-(Z/H)}) \Rightarrow$ *because* : $\ln (x \cdot y) = \ln (x) + \ln(y)$

$\Rightarrow \ln (P_{(z)}) = \ln (P_0) + \ln (e^{-(Z/H)}) \Rightarrow$ *because* : $\ln (e^x) = x \cdot \ln (e) = x \cdot 1 = x$

$\Rightarrow \ln (P_{(z)}) - \ln (P_0) = - \dfrac{z}{H} \Rightarrow$ *because* : $\ln(x) - \ln(y) = \ln(x/y)$

$\Rightarrow z(\text{in meters}) = H \cdot \ln P_0 - H \ln P_{(z)} = H \cdot \ln \left(\dfrac{P_0}{P_{(z)}} \right)$

Formula 1.2: Rearranging air pressure into altitude (*hypsometric equation*).

ments (because of its consistency with the ISA model), is fetched in Formula 1.3. This equation is regularly referred to as *international barometric formula* [17].

$$z(\text{in meters}) = 44330 \cdot \left(1 - \left(\frac{P_{(z)}}{P_0} \right)^{\frac{1}{5.255}} \right) =$$

$$= 44330 - 44330 \left(\frac{1}{101325} \cdot P(z) \right)^{0.1903} =$$

$$= 44330 - 4935.125 \cdot P_{(z)}^{0.1903}$$

Formula 1.3: Rearranging air pressure into altitude (*international barometric formula*).

Nowadays, there is particular research interest in systems applying to barometric altitude readings, which are addressed to determine the absolute vertical distance in between two distinct positions. Those systems rely on the relatively high accuracy of altitude information deriving from MEMS barometric pressure sensors and can be found in an abundance of cutting-edge consumer, medical, and aerospace applications.

There are two different approaches applying to barometric altimetry for the *absolute height* (*h*) determination. The first approach uses a single device (i.e., barometer), which in sequence acquires the air pressure at two different positions (Figure 1.9a,b). In detail, the barometer is initially placed at a particular position for the air pressure acquisition (Figure 1.9a); then it is removed from the initial position and acquires air pressure at a different (higher or lower) location (Figure 1.9b); finally the two pressure measurements are converted to altitude data and the absolute height in between the two positions is determined by the subtraction of the two elevations. Pressure conversion into altitude is performed using either the hypsometric equation (Formula 1.2) or the international barometric formula (Formula 1.3). Since both calculations have a reference to the *standard atmospheric pressure* (P_0) this approach is hereafter referred to as *single-ended altimetry measurement*.

The second approach employs two different barometers which in parallel acquire the atmospheric pressure at two different vertical positions (Figure 1.9c). The pressure measurements are converted to altitude data using either hypsometric equation (Formula 1.4a) or international barometric formula (Formula 1.4b). Since the calculations of Formula 1.4 have no reference to P_0, this approach is hereafter referred to as *differential altimetry measurement*. Such systems require calibration of the barometer devices, which could be possibly performed by an additional measurement at a reference position (Figure 1.9d). Reasons for calibrating barometers are explained later in this chapter. It is worth mentioning that the barometer remaining stable at the position, whether the system runs in calibration or normal mode of operation, is regularly referred to as *base barometer*, while the allied device is also known as a *rover barometer*.

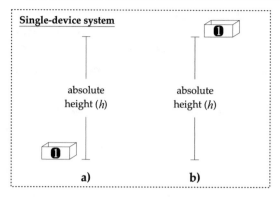
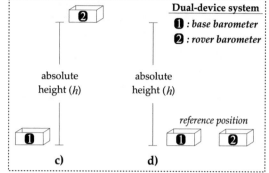

Figure 1.9: Single-ended and differential (barometric) altimetry measurement.

(a) Hyspometric Equation :

$$h = z_A - z_B =$$
$$= \left(H \cdot \ln P_0 - H \ln P_{(zA)} \right) - \left(H \cdot \ln P_0 + H \ln P_{(zB)} \right)$$
$$= H \left(\ln P_{(zB)} - \ln P_{(zA)} \right),$$

where : $z_A > z_B$ and $P_{(zA)} < P_{(zB)}$

(b) International Barometric Formula :

$$h = z_A - z_B =$$
$$= -4935.125 \cdot P_{(zA)}^{0.1903} + 4935.125 \cdot P_{(zB)}^{0.1903} =$$
$$= 4935.125 \cdot \left(P_{(zB)}^{0.1903} - P_{(zA)}^{0.1903} \right)$$

where : $z_A > z_B$ and $P_{(zA)} < P_{(zB)}$

Formula 1.4: Rearranging air pressure into altitude in differential altimetry measurement.

There are particular applications based on either single-ended or differential altimetry measurement systems. Some potential applications of both approaches are presented hereafter.

Figure 1.10 presents an indoor positioning system applying to differential altimetry measurement approach. The system employs three individual barometers attached to a WSN (denoted 1, 2, and 3). Barometer 1 holds the identity of system coordinator, which collects data from the other two sensor devices and makes all three measurements available though the internet, i.e., in terms of *internet of things* (IoT) applications. In this way, users are able to identify their vertical position in a large building (such as a museum) with the use of a particular mobile application. The latter obtains sensor data from the internet and compares them to a measurement acquired by the barometric pressure sensor embedded in the smartphone. When a floor change is identified, the application may automatically send to the user's smartphone specific information related to that particular floor (such as what available exhibits of the museum can be seen on the floor).

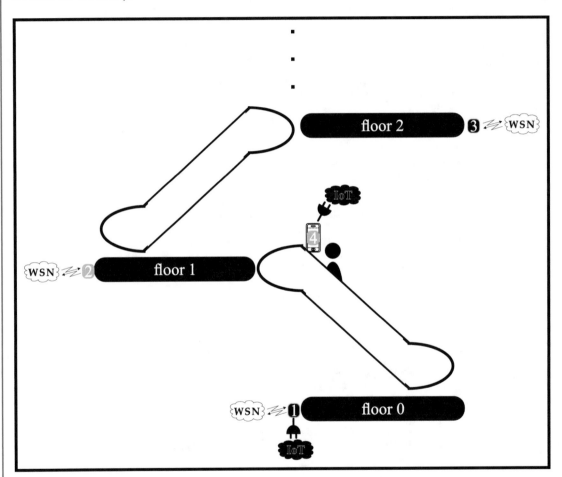

Figure 1.10: Floor identification.

Such a system could be also exploited, along with the use of an *inertial measurement unit* (IMU),[2] for the guidance of visually impaired individuals in indoor environments [18] (such as commercial centers, subway stations, etc.). It should be noted that barometers can be addressed

---

[2]The term IMU refers to an electronic device that measures motion and rotation of a body (and sometimes the magnetic field, as well), through the employment of accelerometers and gyroscopes (and magnometers), respectively. An IMU constitutes the main component of *unmanned aerial vehicles* (UAVs), *inertial navigation systems* (INS), etc.

for the identification of mode of transition among floors, i.e., taking stairs, elevators, or escalators [19]. One of the main reasons barometers have gained ground *in indoor positioning and indoor navigation* (IPIN) systems is due to the inability of the allied *global positioning systems* (GPS) to operate inside buildings because of the satellites' signal loss. Moreover, the GPS receivers derive geometric altitude information, which is of significantly less accuracy compared to the barometric altitude derived by MEMS sensors [16].

At this point it is worth referring to the positive and negative effects on system accuracy when employing either singled-ended or differential altimetry measurement. In single-device systems, accuracy is vulnerable to the atmospheric pressure variations, even for indoor environments. On the other hand, dual-device systems feature a deviation to the sensor's output signal when barometers measure the exact same pressure under identical environmental conditions. This deviation cannot be defined by a single calibration and considerably affects long-term measurement stability [20].

Figure 1.11 depicts the indoor atmospheric pressure graphs acquired by two identical sensors, located at the exact same vertical position. The graphs are addressed to reveal the negative effects of singled-ended and differential altimetry measurement approaches. In regard to the former approach, the atmospheric pressure had been decreased by 0.4 mbar after a one hour period, which is equivalent to an error[3] of 3.2 m. In terms of the latter approach, there is a deviation in sensors' output signal equal to 1.35 mbar (that is, 10.8 m). If we take the time to explore sensors' deviation at small intervals we would reach the interesting conclusion that this deviation is not of constant value and varies unpredictably among different areas of the graph, i.e., it is dependent on the current sensing pressure as well as the existing environmental conditions (mainly, the temperature). Thus, the deviation significantly affects long-term measurement stability and distorts system accuracy.

In order to realize the effects of atmospheric pressure variations on single-device systems we consider the following example: A smartphone application acquires a reference measurement at a particular floor in a building, where each floor height equals to 3 m; after spending one hour on a particular floor the user will be mistakenly notified of a floor transition because of the corresponding change in atmospheric pressure. On the other hand, the effects of sensors' deviation on dual-device systems are quite obvious. In practice, if the sensors are not calibrated in order to identify and remove deviation from the actual measurement, a greater than 10 m error may occur. In addition, if the deviation is identified by one-time calibration a significant error is also possible with the passage of time.

This book addresses all those critical issues present in barometric altimetry systems, through the design of an innovative experimentation model and an in-depth measurement analysis of the acquired data. The book attempts to settle the conditions for a solid theoretical background and a practical guide relevant to systems that apply to barometric altitude readings. Some

---

[3]The 1mbar change in atmospheric pressure equals to a change in altitude of approximately 8 meters at sea level.

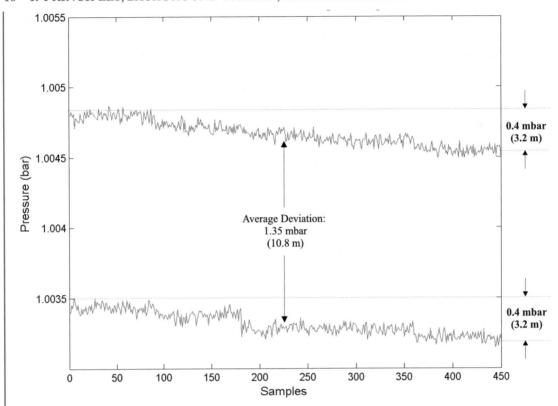

Figure 1.11: Indoor pressure measurements of one hour duration using identical MEMS sensors.

additional examples of potential (contemporary) applications of barometric altimetry are given below.

Figure 1.12 presents a differential altimetry measurement system for avionics applications (exposed to outdoor environment), in consideration of pitch and roll positional angles determination [21]. Roll angle determination is performed through the employment of the dual-device system in the airplane's wings (Figure 1.12a). Pitch angle determination is achieved through the installation of sensors on the front and back of the aircraft (Figure 1.12b). In both cases, a MEMS accelerometer could be addressed for the determination of the aircraft's alignment with the horizon [22], in order to perform re-calibrations of the dual-device system via the identification of the current deviation is sensors' output signal.

Figure 1.13 presents two possible implementations of a dual-device system in healthcare applications. The former (Figure 1.13a,b) evaluates the runner's vertical displacement of the center of mass, which is the key implication for injury mechanics, and it is typically performed through the comparison of video frames [23]. In this setup, the system designer eval-

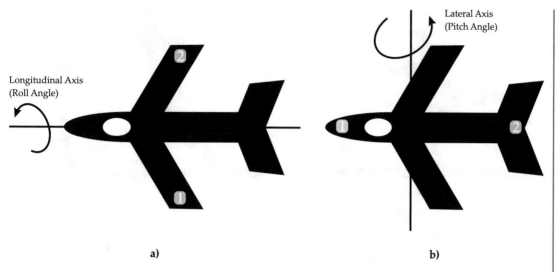

Figure 1.12: Airplane's position angles determination.

uates changes in the output signal of the rover barometer with reference to the base barometer (attached to the treadmill running). The latter (Figure 1.13c,d) is addressed for the detection of the human's fall in terms of a telemedicine system. In this setup, the designer evaluates sharp changes in the deviation of sensors output in case of the human's fall.

Figure 1.14 depicts a single-device system addressed for the evaluation of the vertical jump of a basketball player, in consideration of leisure and sports applications. Due to the short time period expected for this human body motion activity to be completed, atmospheric pressure variations do not affect measurement accuracy at this particular setup. This system requires a rapid acquisition of atmospheric pressure. Therefore, it is recommended to deploy a sensor that provides raw data output during the acquisition process, while performing the time-consuming conversions offline.

There are also particular applications of single-ended altimetry measurement addressed for automotive electronics. For instance, sensor fusion techniques on geometric and barometric altitude information (derived from GPS and MEMS sensor, respectively), are proved to provide more accurate data [24]. In addition, the car alarm can be activated by the change of inside air pressure when detecting a glass breaking or door/window violation.

The aforementioned examples provide an overview of the possibilities of research in MEMS barometers toward vertical position detection. The present effort incorporates all the information needed by scientists and engineers who work on research projects and/or product development of systems applying to barometric altimetry. Recent applications in the literature exploit the MEMS barometric pressure sensors found in smartphones for the implementation of

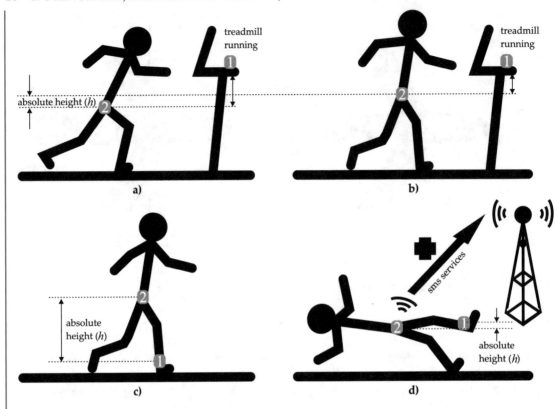

Figure 1.13: Wearable technology for healthcare applications.

position location systems [25, 26]. However, the utilization of the existing mobile technology features particular limitations.

Key features of the book are in line with a low-cost experimentation platform deploying commercial modules, along with low-cost (innovative) characterization techniques that allow experimental observations of particular sensor characteristics. These approaches attempt to provide an effective solution toward low-cost experimentation on MEMS barometric sensors, over the particularly expensive apparatus required for conducting accurate experiments. Such strategies could also be addressed for laboratory training on MEMS pressure sensors and barometric altimetry in the graduate level, while they could be easily adopted by the departments of Electrical and Electronic Engineering and Physics [27].

The next section contributes to the development of wireless barometers and the implementation of a (real-time monitoring) wireless sensor network system. Then, the concluding section provides a deepened understanding to the pressure measurement analysis using Matlab script code. This project aims in extending the prevailing possibilities for research and laboratory

Terrain

Figure 1.14: Wearable technology for leisure and sports applications.

training on MEMS barometers, while also providing to the readers an integrated theoretical and experimental learning model.

REFERENCES

[1] Microelectromechanical systems, https://en.wikipedia.org/wiki/Microelectro mechanical_systems/ [Accessed: Nov-2016].

[2] W. P. Eaton and J. H. Smith, Micromachined pressure sensors: Review and recent developments, *Smart Materials and Structures*, 6(5), pp. 530–539, 1997. DOI: 10.1088/0964-1726/6/5/004.

[3] A. Saxena, M. M. Singh and V. Singh, The state of art of MEMS in automation industries, in *Proc. of the Innovative Trends in Applied Physical, Chemical, Mathematical Sciences, and Emerging Energy Technology for Sustainable Development (APCMET'14)*, New Delhi, pp. 1–5, 2014.

[4] CoventorWare, http://www.coventor.com/mems-solutions/products/coventorware/ [Accessed: Nov-2016].

[5] About X-FAB, http://www.xfab.com/about-x-fab/ [Accessed: Nov-2016].

[6] ASE Group, http://www.aseglobal.com/ [Accessed: Nov-2016].

[7] J. Fraden, *Handbook of Modern Sensors: Physics, Designs and Applications*, Springer-Verlag, NY, 2004.

[8] *Pressure Transducers: Technical Note 1*, Sensata, 2008.

[9] E. Gassmann, Pressure sensors fundamentals: Interpreting accuracy and error, *Chemical Engineering Progress*, 92(1), pp. 37–45, 2014.

[10] J. Sanders, Why temperature compensation really matters for pressure measurement. http://www.additel.com/UploadFiles/WhitePaper/Why-Temperature-Compensation-Really-Matters-for-Pressure-Measurement.pdf [Accessed: Nov-2016].

[11] *BME280 Combined Humidity and Pressure Sensor*, Bosch Sensortec, Germany, 2015.

[12] *LPS25HB MEMS Pressure Sensor: 260–1260 hPa Absolute Digital Output Barometer*, STMicroelectronics, 2015.

[13] S. Katsikas, D. Dimas, A. Defigos, A. Routzomanis, and K. Mermikli, Wireless modular system for vessel engines monitoring, condition based maintenance and vessel's performance analysis, in *Proc. of the 2nd European Conference of the Prognostics and Health Management Society (PHME'14)*, Nantes, France, pp. 1–10, 2014.

[14] K. J. Rebello, Applications of MEMS in surgery, *IEEE Proceedings*, 92(1), pp. 43–55, 2004. DOI: 10.1109/jproc.2003.820536.

[15] *Atmospheric Pressure Measurements: Technical Note*, Honeywell, 2004.

[16] Use of barometric altitude and geometric altitude information in ADS-B message for ATC applications, in *Proc. 8th Meeting of the South East Asia and Bay of Bengal Sub-regional ADS-B Implementation Working Group*, Yangon, Myanmar, pp. 1–4, 2012.

[17] *BMP180 Digital Pressure Sensor*, Bosch Sensortec, Germany, 2013.

[18] J. Zegarra Flores and R. Farcy, Indoor navigation system for the visually impaired using one inertial measurement unit (IMU) and barometer to guide in the subway stations and commercial centers, in *Computers Helping People with Special Needs*, Springer, Paris, France, pp. 411–418, 2014. DOI: 10.1007/978-3-319-08596-8_63.

[19] S. Vanini, F. Faraci, A. Ferrari, and S. Giordano, Using barometric pressure data to recognize vertical displacement activities on smartphones, *Computer Communications*, 87, pp. 37–48, 2016. DOI: 10.1016/j.comcom.2016.02.011.

[20] D. E. Bolanakis, K. T. Kotsis, and T. Laopoulos, A prototype wireless sensor network system for a comparative evaluation of differential and absolute barometric altimetry, *IEEE Aerospace and Electronic Systems Magazine*, 30(11), pp. 20–28, 2015. DOI: 10.1109/maes.2015.150013.

[21] P. Paces, J. Popelka, and T. Levora, Advanced display and position angles measurement systems. In *Proc. of the 28th Congress of the International Council of the Aeronautical Sciences, ICAS*, Brisbane, Australia, P6.3.1–P6.3.14, 2012.

[22] S. Luczak, Single-axis tilt measurements realized by means of MEMS accelerometers, *Engineering Mechanics*, 18, pp. 341–351, 2011.

[23] R. Souza, An evidence-based videotaped running biomechanics analysis, *Phys. Med. Rehabil. Cli N. Am.*, 27(1), pp. 217–236, 2016. DOI: 10.1016/j.pmr.2015.08.006.

[24] V. Zaliva and F. Franchetti, Barometric and GPS altitude sensor fusion, in *Proc. IEEE International Conference on Acoustics, Speech and Signal Processing*, Florence, pp. 7575–7579, 2014. DOI: 10.1109/icassp.2014.6855063.

[25] H. Xia, X. Wang, Y. Qiao, J. Jian, and Y. Chang, Using multiple barometers to detect floor location of smart phones with built-in barometric sensors for indoor positioning, *Sensors*, 15, pp. 7857–7877, 2015. DOI: 10.3390/s150407857.

[26] K. Muralidharan, A. J. Khan, A. Misra, R. K. Balan, and S. Agarwal, Barometric phone sensors—More hype than hope!, in *Proc. of ACM Hotmobile*, Santa Barbara, CA, pp. 1–6, 2014. DOI: 10.1145/2565585.2565596.

[27] D. E. Bolanakis, K. T. Kotsis, and T. Laopoulos, Ethernet and pc-based experiments on barometric altimetry using MEMS in a wireless sensor network, *Computer Application in Engineering Education*, 24(3), pp. 428–442, 2016. DOI: 10.1002/cae.21722.

CHAPTER 2

System Prototyping

This chapter contributes to the development of wireless barometers and the implementation of a real-time monitoring wireless sensor network (WSN) system.

2.1 SINGLE-ENDED ALTIMETRY MEASUREMENT SETUP

Figure 2.1 presents the architecture of a system of a single-ended altimetry measurement method. The proposed implementation applies to the IEEE 802.15.4 packet data protocol for 2.4 GHz wireless communications. This technical standard sustains the arrangement of wireless *personal areas networks* (PANs) of low cost, power, and rate, toward a reliable communication among nearby devices, a particularly effective solution in noisy environments. The IEEE 802.15.4 standard constitutes the basis for many protocol stacks, including the familiar ZigBee stack.

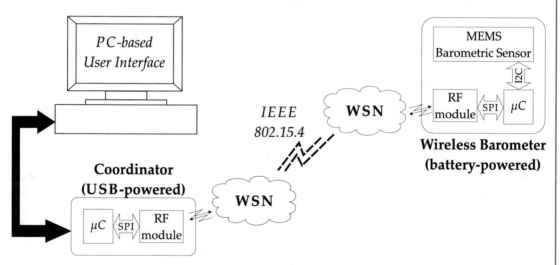

Figure 2.1: System architecture of single-ended altimetry measurement method.

IEEE 802.15.4 networks can be of either *peer-to-peer* or *star* topology, while each network may employ two different kinds of nodes, that is, the *full-function devices* (FFDs) and the *reduced-function devices* (RFDs). The former are the ones that work as network coordinators, but they can also work as common nodes. The latter may only communicate with the network's FFDs.

Because of the demand of a simplified implementation, the proposed system employs FFDs in a star topology, where one of them constitutes the network coordinator and the rest of them are the battery-operated measurement devices. Thereby, the same driver (i.e., the underlying firmware code) can be employed by each particular device of the system, configuring them as PAN coordinators.

The (USB-operated) network coordinator of the proposed setup receives commands from a graphical user interface (GUI) running on a host PC, and initiates a wireless transaction with the measurement device (aka *barometer*). The barometer obtains a measurement of the existing air pressure and temperature sensed by the MEMS device, and sends data back to the coordinator in ASCII format. As soon as the coordinator receives measurements from the barometer device, it forwards data to the GUI running on the host PC. The brain of the system coordinator and barometer devices is a microcontroller unit which apart from the wireless interface controls the communication with the USB bus of the host PC (in the former device), as well as the communication with the MEMS sensor (in the latter device).

Figures 2.2 and 2.3 present, respectively, the prototype hardware of the barometer and coordinator device. The barometer consists of the following five *printed circuit boards* (PCBs):

1. **Arduino Uno** [1]: the microcontroller board employing ATmega328 device [2];

2. **Adafruit PowerBoost Shield** [3]: lithium-ion and lithium polymer battery charger;

3. **Arduino Uno Click Shield** [4]: an extension for Arduino Uno board to support connectivity of Mikroelektronika's click boards;

4. **Bee Click** [5]: module featuring mrf24j40ma [6] IEEE 802.15.4 radio transceiver module from Microchip; and

5. **Weather Click** [7]: module featuring bme280 [8] MEMS sensor from Bosch Sensortec.

The same PCBs are addressed for the implementation of the network coordinator. The latter device employs PCBs 1, 3, and 4 as depicted by Figure 2.3. It is worth noting that installation of both coordinator and barometer devices rely on merely a plug of the add-on mezzanine boards, onto the Arduino Uno (i.e., no additional hardware occupation is required). Current consumption for the battery-operated measurement device is given in Table 2.1. The overall consumption of current was (experimentally) found to be of approximately 150 mA.

The Adafruit PowerBoost 500 shield employs a red light–emitting diode (LED), named LOW, which is turned on when the battery voltage reaches (or drops below) 3.2 V. Accordingly, employing in the system a battery of 3.7 V nominal voltage which is regularly charged up to 4.2 V, it will be discharged approximately 25% of the maximum capacity with this particular setup (i.e., 4.2 V–4.2 Vx25%$\cong$ 3.2 V). Thereby, a battery of 3,600 mAh capacity is able to supply the system for six successive hours; that is, 3,600 mAh x 25%= 900 mAh $\implies$ 900 mAh÷150 mA = 6 h.

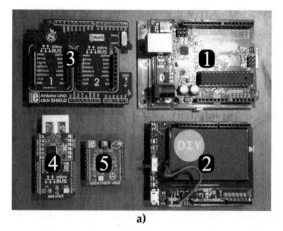
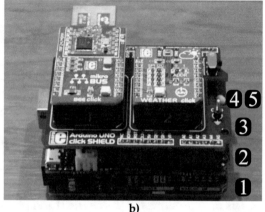

a)   b)

Figure 2.2: Barometer device (battery-powered device).

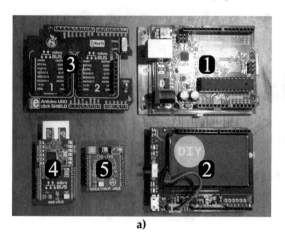
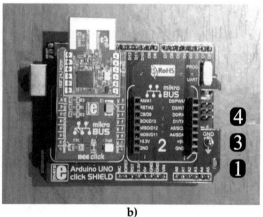

a)   b)

Figure 2.3: Coordinator device (USB-powered).

Figure 2.4 presents the interface protocol developed for the coordinator and barometer device (depicted in black and grey color, respectively). The process starts with a command sent from the GUI running on host PC,[1] and which is subsequently received by the network coordinator of the proposed system. As soon as the command is received by the coordinator, the latter transmits wirelessly the unique address of the barometer attached to the network (Timeframe1). Then the coordinator is inserted into a state of waiting data from the barometer device (Timeframes:2–4). The measurement device, which is constantly waiting for incoming data (Timeframe1), identi-

---

[1]The overall process is synchronized by the GUI (described later in this chapter) which sequentially sends through the USB port the *"Obtain Measurements"* command to the network coordinator, and then receives pressure and temperature measurements from the latter device.

Table 2.1: Consumption of the battery-powered barometer

Device	Current Consumption
Arduino Uno	64.4 mA
Adafruit PowerBoost 500 Shield	3.6 mA
Arduino Uno Click Shield	0 mA
Bee Click	78 mA
Weather Click	7 mA
Barometer	**153 mA**

fies its unique address (Timeframe2) and obtains pressure and temperature measurements from the MEMS sensor, through an I2C interface (Timeframe3). Then it sends data back to the co-ordinator (Timeframe 4). Finally, the coordinator obtains measurement data from the wireless barometer and the latter is inserted into the state of waiting for wireless data (Timeframe 5). The overall process is repeated from the beginning (Timeframe1) and the number of iterations depends on the corresponding configuration of GUI running on the host PC.

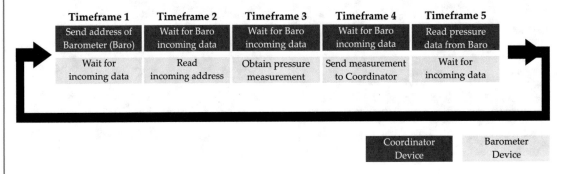

Figure 2.4: Pressure and temperature data acquisition protocol (single-ended altimetry).

2.2   DIFFERENTIAL ALTIMETRY MEASUREMENT SETUP

Figure 2.5 presents the architecture of a system of a differential altimetry measurement method. The system has been designed so that measurement devices can be easily integrated into the existing setup, by changing merely the effective (unique) address of each barometer device. The proposed system in the current implementation (i.e., firmware and software) admits up to four unique barometers.

The setup of the differential altimetry measurement method (Figure 2.5) can be addressed for single altimetry as well, through an analysis of the measurements obtained from the rover barometer only. The interface protocol of differential altimetry setup is given in Figure 2.6.

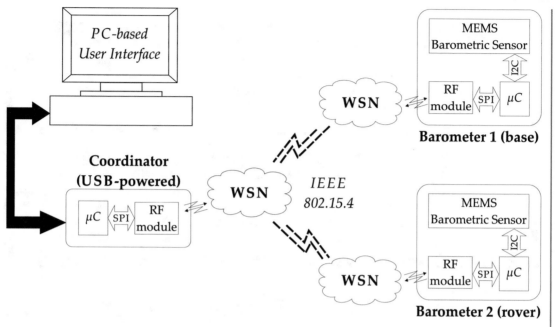

Figure 2.5: System architecture of the differential altimetry measurement method.

The protocol ensures that measurement data are simultaneously obtained from both measurement devices. In detail, as soon as the coordinator receives the corresponding command from the host PC, it transmits the address of barometer 1, while barometers wait for incoming data (Timeframe1). Both barometers receive the wireless data (Timeframe2) and acquire a measurement from the available MEMS sensor (Timeframe3). This is because the effective address of the first measurement device in the network is associated with the *"obtain measurement"* command and, hence, measurements are concurrently obtained from all measurement devices of the wireless sensor network. Thereafter, the barometer whose address is in agreement with the received data transmits back to the coordinator the acquired measurement data, while the allied barometer enters back into the state of waiting for incoming data (Timeframe 4). As soon as the coordinator receives measurements from barometer 1 (Timeframe 5), it transmits the effective address of barometer 2 (Timeframe 6). The latter identifies its effective address (Timeframe 7) and transmits back to the coordinator the measurements data (Timeframe 8), which were previously obtained in Timeframe 3. The coordinator receives measurements from barometer 2 and both barometers enter into *"Wait for incoming data"* state (Timeframe 9). The overall process is repeated from Timeframe1 in case the host PC initiates a new measurement.

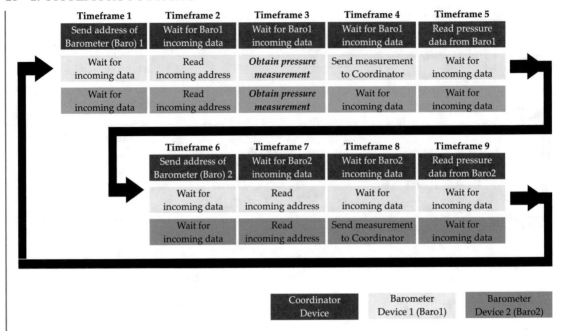

Figure 2.6: Pressure and temperature data acquisition protocol (differential altimetry).

As mentioned earlier, the proposed system in the current implementation is able to obtain measurements from one, two, three, or four[2] different barometers. This configuration is decided by the GUI running on the host PC. The interface protocol is similar to the one described above. In detail, when the network coordinator transmits address 1, all barometers are forced to obtain measurement from the MEMS sensor employed by the device (Figure 2.7a). Then the barometer associated with this particular address sends measurements back to the coordinator (Figure 2.7b). The coordinator transmits address 2 in order to receive measurements from the second—in the system—barometer (Figure 2.7c). Barometer 2 responds to this call and transmits back to the coordinator the measurements (Figure 2.7b). The same procedure is repeated for the barometers of effective address 3 and 4.

The firmware among the different barometer devices is common, although a declaration of the effective address of each barometer device is required. The designer should use values from one to four according to the number of devices employed by the system.

The flowchart of the barometer's firmware code is given in Figure 2.8. The code starts with the requisite configurations of the MEMS sensor device and, thereafter, the microcontroller waits for incoming data to be received by the radio transceiver module. Upon reception of the wireless data, the rf module notifies the microcontroller through an output pin, while the latter

---

[2]The maximum number of the employed measurement devices can be further increased with minor revisions in firmware and software.

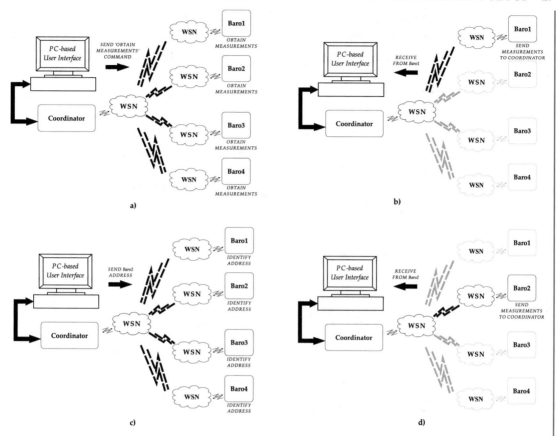

Figure 2.7: Data acquisition through four barometers.

device evaluates if the incoming data are equal to the *"obtain measurement"* command. If the condition is confirmed, the microcontroller acquires pressure and temperature measurements from the MEMS sensor device and, thereafter, it converts them to ASCII format (in the form given by the figure). Subsequently to this step, or in case the evaluated condition is false, the microcontroller explores if the incoming data match to the effective address of the barometer device. At this point of the code, one of the following conditions can be in effect:

1. The *"obtain measurement"* command was received by the first barometer in the WSN and, hence, the incoming data match with the effective address of the device, which thereby transmits the acquired measurements back to the coordinator.

2. The *"obtain measurement"* command was received by a barometer of effective address other than 1 and, therefore, the device acquires measurements and then the process starts from the beginning.

3. An address other than 1 was received by the barometer and the latter device evaluates if this value is equivalent to the barometer's effective address. If the condition is true, the device sends to the coordinator the measurement data acquired by a previous run of the firmware process; otherwise, the program code starts over again (i.e., the coordinator asked for data from a different barometer).

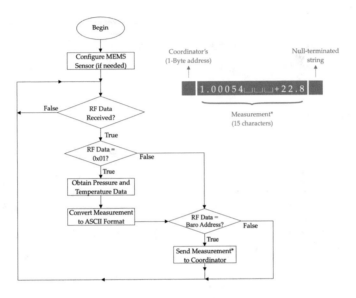

Figure 2.8: Flowchart of barometer device.

The flowchart of the coordinator's firmware code is given in Figure 2.9. The microcontroller performs the necessary configurations in order to establish the communication with the USB port of the host PC and, thereafter, it waits to receive a command from PC's GUI. As soon as the coordinator receives the *"obtain measurement"* command from the USB port, it forwards the same request to the barometers which constantly wait to receive wireless data. As mentioned earlier in the chapter, this command is equivalent to the effective address of barometer 1 and, hence, the coordinator enters the state of receiving wireless data from this particular barometer, as soon as the transmission ends.

It is worth mentioning that the coordinator's firmware utilizes a timeout counter, which restarts the process from the beginning in case barometers do not respond to the coordinator's calls. In case the communication between the coordinator and wireless barometers is successfully established, the former sequentially obtains measurement data of each barometer device. The coordinator appends a comma (,) character in between measurements of different barometers, as well as a null (\0) character at the end of the final measurement. When the network coordinator receives measurements from all barometers attached to the WSN, it sends data to the USB port of the host PC.

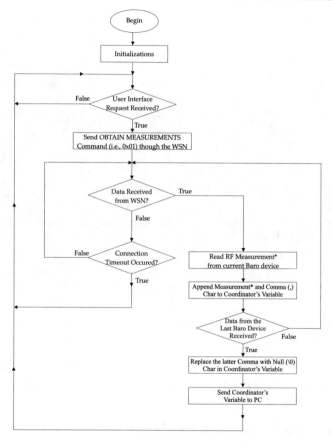

Figure 2.9: Flowchart of coordinator device.

2.3   SOFTWARE INTERFACING

The GUI running on the host PC is developed in the LabVIEW environment using C functions and call-to-function rules. The LabVIEW-based GUI, depicted in Figure 2.10, synchronizes the communication of the host PC with the network coordinator through a USB port and receives pressure and temperature measurement data from the wireless barometers. Data are stored to a text file for a further, offline analysis in Matlab.

The control buttons and indicators of LabVIEW GUI are as follows:

- **START**: initiates the data acquisition process.

- **USB COM Port**: selects the communication (usb-to-serial) port at which the Arduino Uno board is attached.

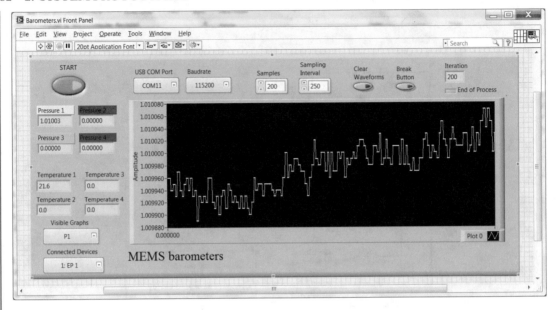

Figure 2.10: LabVIEW-based user interface (one device connected).

- **Baudrate**: selects the serial data speed between the host PC and Arduino; this selection should be identical to the corresponding configuration of the firmware running on the network coordinator.

- **Samples**: determines the number of samples that will be acquired by the system.

- **Sampling Interval**: determines the time interval among successive measurements (minimum value is defined by the coordinator's firmware).

- **Clear Waveforms**: clears measurements from the GUI screen.

- **Break Button**: stops the existing measurement procedure.

- **Iteration**: enumerates the current sample of measurement acquired by the system.

- **Pressure1–4**: reveals the existing barometric pressure acquired by the corresponding sensor device (i.e., Sensor1–4).

- **Temperature1–4**: reveals the existing temperature value acquired by the corresponding sensor device (i.e., Sensor1–4).

- **Connected Devices**: selects the number of barometer devices employed by the WSN (i.e., from 1–4 for this particular implementation); in the example of Figure 2.10 only one barometer device is assumed to be attached to the system.

- **Visible Graphs**: selects the number of visible graphs that will appear on screen; for instance, if the system employs (4) barometers and this control button is set to (2), only graphs obtained from Sensor 1 and 2 will appear on screen.

- **Screen**: illustrates the pressure graphs obtained from the barometer devices (according to the initialization of *Visible Graphs* option).

2.4   ARRANGEMENT OF LOW-COST EXPERIMENTAL OBSERVATIONS

The particular expensive apparatus required for conducting accurate experiments on MEMS pressure and barometric pressure sensors has been explored, in detail, in the introductory chapter of the book. However, some experimental observations on the MEMS sensors' specifications are necessary in order to proceed to an effective design of a system applying to vertical position location detection. For that reason, an airtight enclosure has been selected to accompany the experimental observations performed later in the book.

The selected airtight enclosure is presented in Figure 2.11. A handheld *vacuum pump* at the top of the enclosure renders feasible the air abstraction and hence, the generation of vacuum. The wireless feature of barometer devices allows us to insert measurement devices inside of the box and acquire samples at different pressure levels. At first, we generate a pressure level below the existing atmospheric pressure, but not below the lower pressure level of the sensor's operating range (for bme280 the operating range is equal to 0.3–1.1 bar [8]). Then we apply a gentle

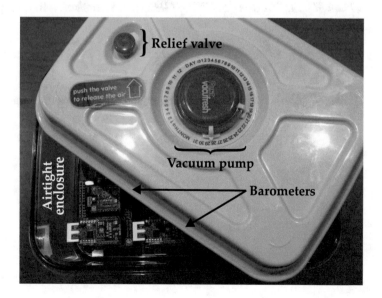

Figure 2.11: Setup for the generation of vacuum.

mechanical force to the relief valve at the top of the enclosure. The latter action increases slightly[3] the pressure level inside of the enclosure.

Because the ambient air pressure will never fall much lower than the 1bar, we could make experimental observations at pressure levels nearby to the sensing atmospheric pressure, e.g., from 0.95–0.99 bar. Observation in sensors' output signal over the existing atmospheric pressure is not feasible with this setup. However, we could reach some interesting conclusions in consideration of the deviation in sensors' output signal (at different pressure levels) when equivalent sensors acquire the exact same pressure under identical temperature and humidity conditions. Such issues are explored in the following chapter.

REFERENCES

[1] Arduino UNO and Genuino UNO, `https://www.arduino.cc/en/Main/arduinoBoardUno` [Accessed: Feb-2017].

[2] *ATmega48PA/88PA/168PA/328P*, Atmel Corporation, Oct. 2009.

[3] *Adafruit PowerBoost 500 Shield*, Adafruit Industries, Aug. 2016.

[4] Arduino Uno Click Shield, `https://www.mikroe.com/shields/` [Accessed: Feb-2017].

[5] Bee Click, `https://www.mikroe.com/products/#add-on-boards`

[6] *MRF24J40 Data Sheet IEEE 802.15.4™ 2.4 GHz RF Transceiver*, Microchip Technology Inc., Aug. 2010.

[7] Weather Click, `https://www.mikroe.com/products/#add-on-boards` [Accessed: Feb-2017].

[8] *BME280 Combined Humidity and Pressure Sensor*, Bosch Sensortech, Oct. 2015.

---

[3]The smallest increment is experimentally measured approximately 100 μbar with this particular setup.

CHAPTER 3

Effects on Measurement Accuracy

This chapter provides a deepened understanding of pressure measurement analysis. The analysis is performed in Matlab using script code, which can be found in the Appendix.

3.1 GAUSSIAN WHITE NOISE SIGNAL IN BAROMETRIC SENSOR OUTPUT SIGNAL

In the introductory chapter of the book we explored the reasons why differential altimetry measurement is of increased accuracy over single-ended altimetry. Because ambient air pressure is simultaneously applied to both sensors of the system, the measurement result is less vulnerable to the possible variations of atmospheric pressure. While this measurement method may be interpreted as the solution over the problems inherent in single-ended altimetry, it features the deviation in sensors' output signal (when devices measure the exact same pressure value) which affects long-term measurement stability.

Before proceeding to the analysis of vertical position determination based on these two particular measurement methods (that is, differential and single-ended altimetry), we explore the deviation in sensors' output signal and the main factors that affect measurement accuracy. The low-cost experimentation methods presented hereafter could be found very assistive toward an improved design of differential positioning systems, addressed to evaluate absolute height differences. Failing to attend to such issues could lead to bad system design, and the poor design may worsen accuracy of differential altimetry compared to single-ended altimetry (though the opposite scenario would generally be expected).

The experiments presented hereafter apply to BME280 [1] sensor of Bosch Sensortec MEMS vendor, but they could straightforwardly be addressed for any available sensor device. In order to understand the measurement analysis presented later in the chapter, we should keep in mind that if we explore the output of a MEMS barometric sensor at a constant pressure level we will obtain a Gaussian white noise signal. Unfortunately, this observation is only feasible with the employment of a pressure controller, able to provide a pressure level of invariable value. When a barometric sensor acquires the ambient air pressure, it senses the pressure variations as well (even if the data acquisition process is performed in an indoor environment, where pressure variations are somehow limited).

The Gaussian white noise signal in a sensor's output can only be observed for a small time interval, at which the atmospheric air pressure does not have the time to change. Unfortunately, the usual sampling rate of an embedded sensing system (that is, a few hundreds of msec) together with the regular time needed for (indoor) atmospheric pressure to be increased/decreased, does not allow us to acquire an adequate number of samples (and the 30-sample rule of thumb is not always appropriate). In order to extend the time of keeping atmospheric pressure at a constant level (without the use of a pressure controller) we may address the airtight enclosure presented in Chapter 2. This particular setup renders feasible the acquisition of more samples at a constant pressure level, before a change in the temperature of the closed (inside) air would cause a proportional change in the sensing pressure.

To clarify this particular process, Figure 3.1 presents 125 pressure samples from measurements obtained by two barometer devices located inside of an airtight enclosure, at a pressure level approximately equal to 0.985 bar. The latter value is estimated by the sensing output of

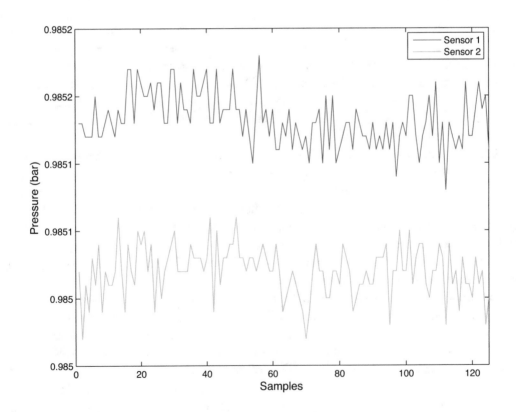

Figure 3.1: Pressure measurements inside of the airtight enclosure (samples 1–125).

barometers. As we observe in this particular figure, the pressure variations inside of the enclosure are limited. If we focus our attention in a narrower range of pressure samples (for instance, for the samples 66–125 depicted in Figure 3.2), we observe that those variations are further eliminated. However, the deviation in sensors' output signal for the overall set (presented in Figure 3.3) is somewhat constant, as the reduced pressure variations inside of the enclosure simultaneously affect both sensors of the system.

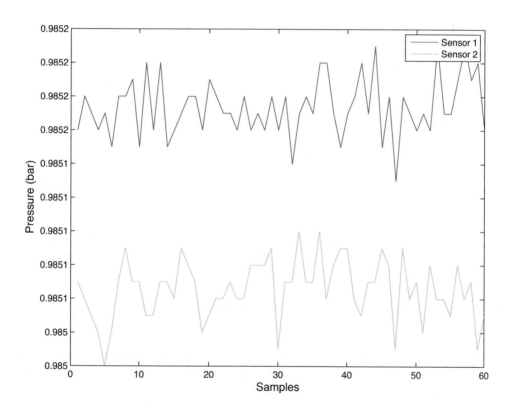

Figure 3.2: Pressure measurements inside of the airtight enclosure (samples 66–125).

Subsequently, the histogram representing distribution of the pressure deviation dataset is normalized over the *sample mean* value $x = 112$ ubar, and characterized by the *sample standard deviation* $s = 23$ ubar (Figure 3.4). The latter value constitutes the *RMS Noise* of pressure deviation in sensors' output signal. To convert RMS into *Peak-to-Peak Noise* for the 99.7% of the overall dataset we need six standard deviations and hence, ppNoise $\cong 140$ ubar in this particular example.

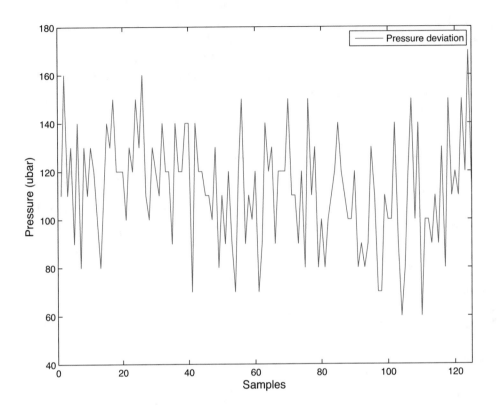

Figure 3.3: Pressure deviation of the two barometers' output signal.

Research endeavors in the literature, which apply to barometric altimetry, have identified that deviation in the output signal of two barometric sensors varies over time [2] and affects long-term measurement stability [3]. Thereby, researchers often consider differential altimetry accurate for altitude measurements in the sense of short-term stability [4]. While this deviation varies randomly and can only be observed experimentally, there are particular methods that can be addressed to sustain long-term measurement stability of the system under particular circumstances.

The examples presented hereafter aim at exploring the most important effects on system accuracy in differential altimetry measurements. Of all possible effects, the temperature influence is in the top of the hierarchy. Another effect on system accuracy, which is worth exploring, is influenced by the existing pressure acquired by the sensor devices. This is possibly the most important factor of deviation randomness in the sensors' output signal, arising from dissimilarities in the sensors' pressure-response curves [5]. From the following two examples of

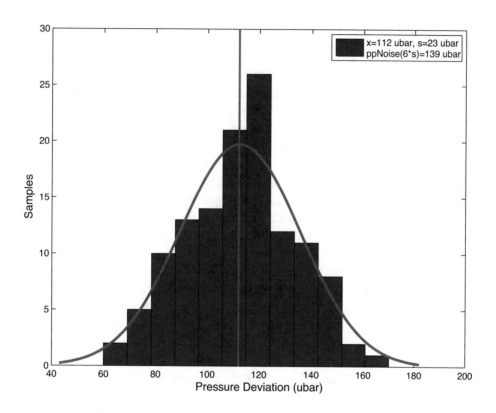

Figure 3.4: Histogram of pressure deviation in sensor's output (normalized distribution).

measurement analysis, we are able to determine sample mean (hereafter denoted x) and sample standard deviation (hereafter denoted s) of each particular sampling distribution. In addition, we can intuitively understand that the estimator is biased (i.e., overestimating or underestimating true value of sensors' deviation), although bias cannot be assessed as the population mean value (hereafter denoted μ) is not known.

3.2 TEMPERATURE INFLUENCES ON BAROMETRIC ALTITUDE MEASUREMENTS

To evaluate the temperature influence on measurement accuracy of a differential altimetry setup, we run the following experiment:

Inside of the airtight enclosure we place the two wireless barometer devices (presented in the previous chapter), which hold BME280 sensors. The enclosure is addressed for reducing the pressure variations sensed by the two devices, as no vacuum is generated for this particular example. The overall setup is placed inside of an oven which has been previously warmed up to 60°C. When temperature of barometers reaches this temperature level, the devices are moved outside the oven and the data acquisition process is initiated; the data acquisition is constantly performed as the temperature of barometers is cooled down to the level of room temperature, which is about 15°C for this particular example. Afterward, the barometers are placed inside of a refrigerator's deep freezer till their temperature reach the value of 0°C; then, the devices are moved to refrigerator's regular freezer and the data acquisition process starts over again, till the devices reach temperature level of 7°C; finally, the devices are moved outside the fridge and the data acquisition process continues till the temperature level of 15°C.

With this particular experiment we are able to capture a large amount of pressure samples and evaluate deviation in sensors' output, for the temperature range 0–60°C. It is worth noting the operating temperature range of BME280 sensor at full accuracy which, according to the sensor's datasheet, is equal to 0–65°C. Figure 3.5 depicts 5,350 pressure samples acquired for the temperature range 0–60°C. More accurate measurements would be possible with the employment of a pressure controller along with an environmental chamber (i.e., a setup of hundred thousand dollars), where the sensors' output signals would be kept constant at a desired pressure level. Due to the absence of such an apparatus, the pressure measurements depicted in Figure 3.5 vary. More precisely we observe an increment in the acquired pressure of approximately 1mbar. This increment is due to the fact that temperature in this particular example decreases from 60°C–0°C, and the pressure response is inversely proportional to temperature (except the condition of a closed air inside of a chamber, where an increment in temperature causes the inside air pressure to be increased, and vice versa).

The measurements depicted in Figure 3.6 illustrate that deviation in the output signal of two BME280 sensors significantly increases at high temperature ranges. If we calculate and plot the absolute difference in sensors' output signal we observe that deviation varies 0.5 mbar for the overall temperature range. Consequently, if we don't take into account temperature influence we might obtain measurement error equivalent to 4 m (at sea level).[1]

The measurement analysis is separated into different areas relative to the temperature ranges where deviation remains somewhat steady and hence, we are expecting a normalized sampling distribution. Figure 3.7 presents pressure measurements of the two barometers at the temperature range 60–50°C. The plot of the corresponding deviation at this particular temperature range is given in Figure 3.8, while the corresponding histogram of pressure deviation is depicted in Figure 3.9. The title of the latter figure incorporates sample mean and peak-to-peak noise values.

---

[1]As a reminder, a change of 1 mbar in the atmospheric pressure equals a change in altitude of approximately 8 meters at sea level.

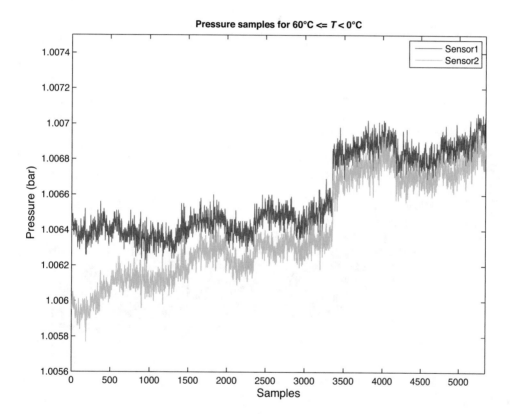

Figure 3.5: Sensors' output signal for the temperature range 60–0°C.

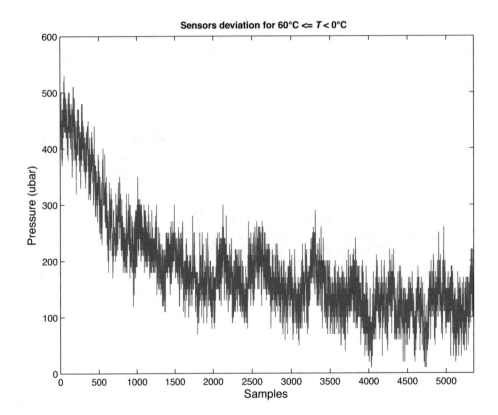

Figure 3.6: Deviation in sensor's output signal for the temperature range 60–0°C.

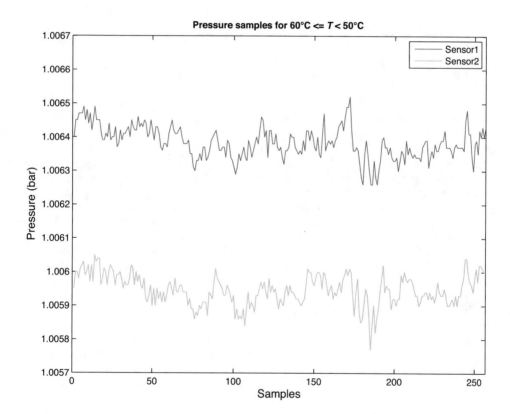

Figure 3.7: Pressure measurements for the temperature range 60–50°C.

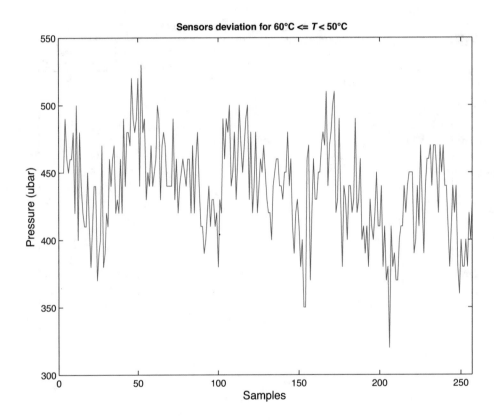

Figure 3.8: Deviation for the temperature range 60–50°C.

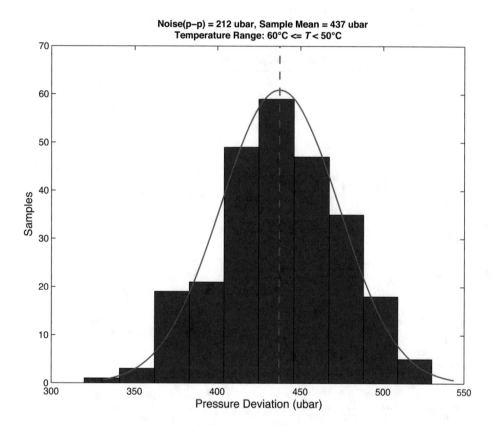

Figure 3.9: Histogram of pressure deviation for the temperature range 60–50°C.

The same process is repeated for the temperature ranges:

1. 50–40°C (Figures 3.10–3.12);

2. 40–30°C (Figures 3.13–3.15);

3. 30–15°C (Figures 3.16–3.18);

4. 15–0°C (Figures 3.19–3.21).

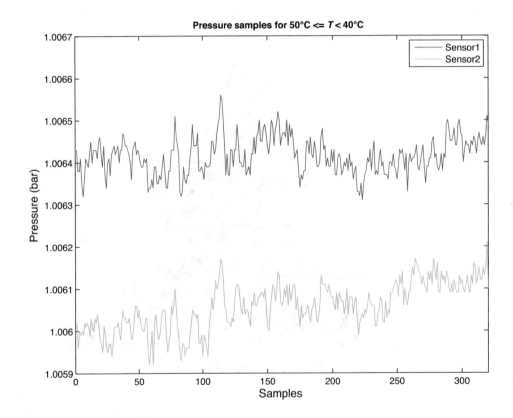

Figure 3.10: Pressure measurements for the temperature range 50–40°C.

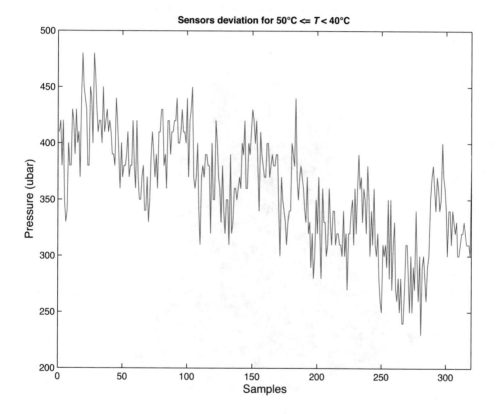

Figure 3.11: Deviation for the temperature range 50–40°C.

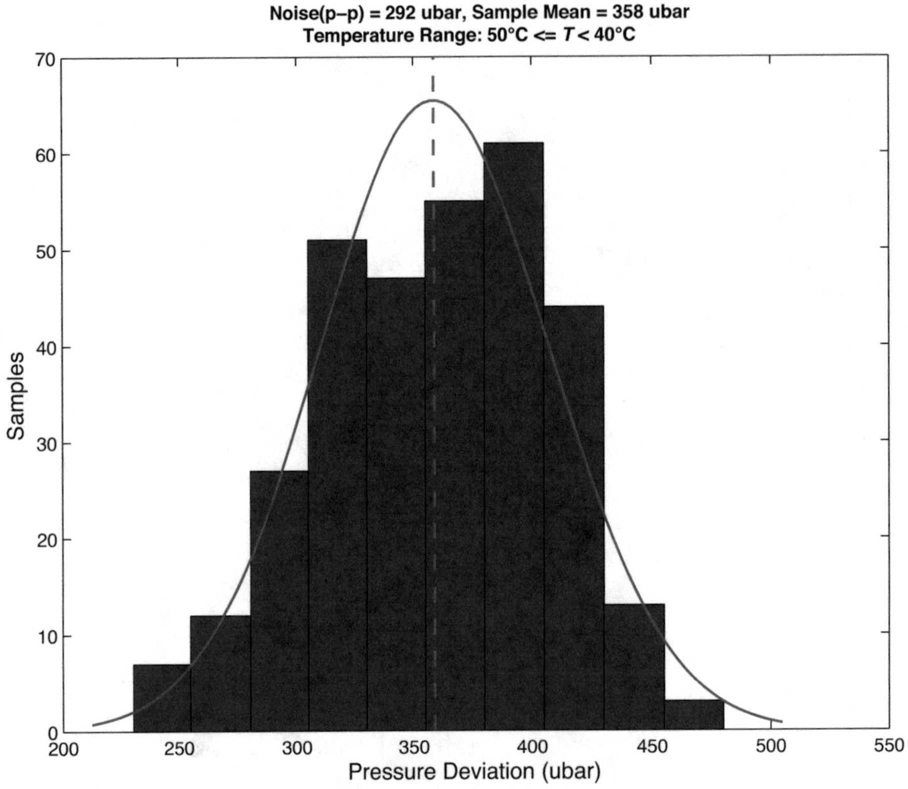

Figure 3.12: Histogram of pressure deviation for the temperature range 50–40°C.

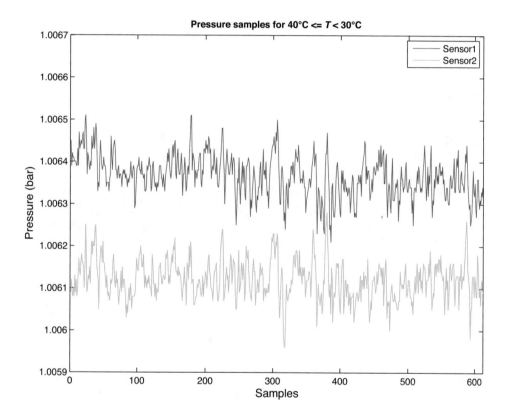

Figure 3.13: Pressure measurements for the temperature range 40–30°C.

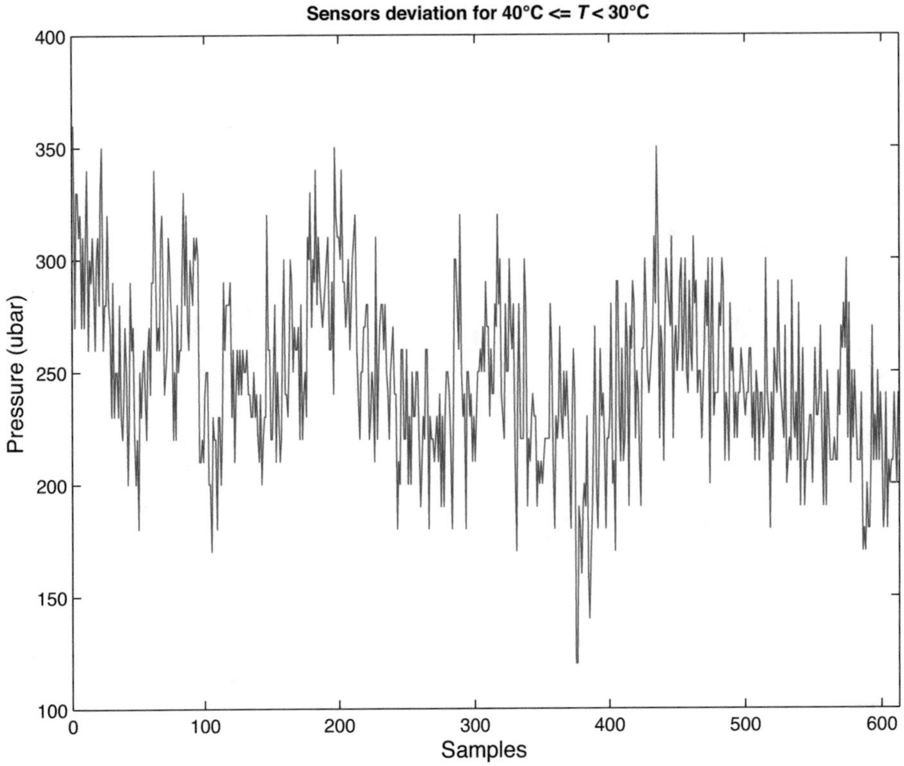

Figure 3.14: Deviation for the temperature range 40–30°C.

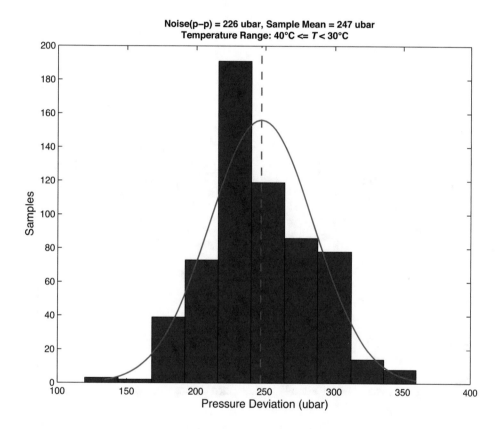

Figure 3.15: Histogram of pressure deviation for the temperature range 40–30°C.

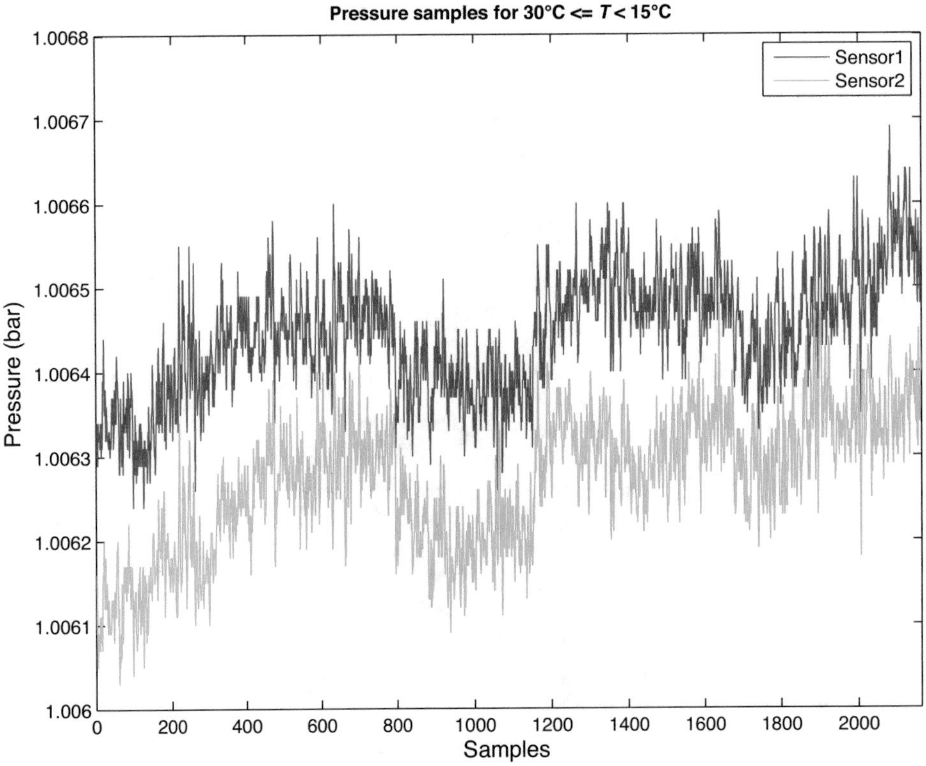

Figure 3.16: Pressure measurements for the temperature range 30–15°C.

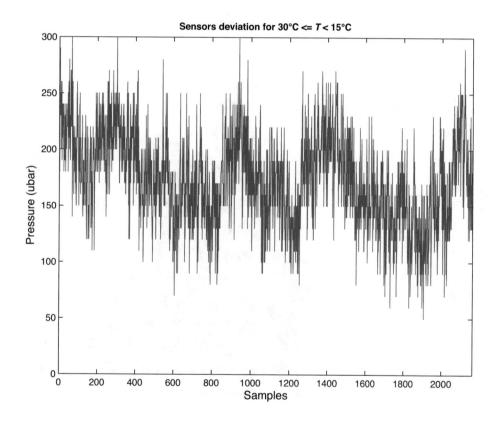

Figure 3.17: Deviation for the temperature range 30–15°C.

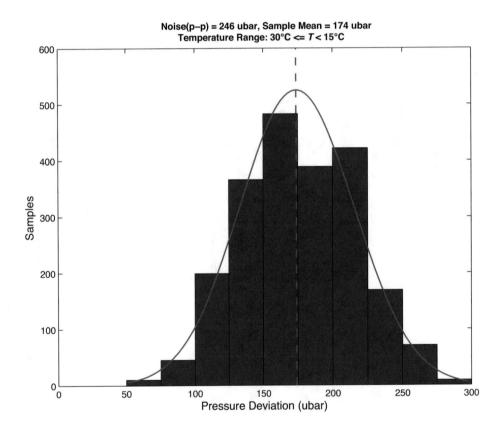

Figure 3.18: Histogram of pressure deviation for the temperature range 30–15°C.

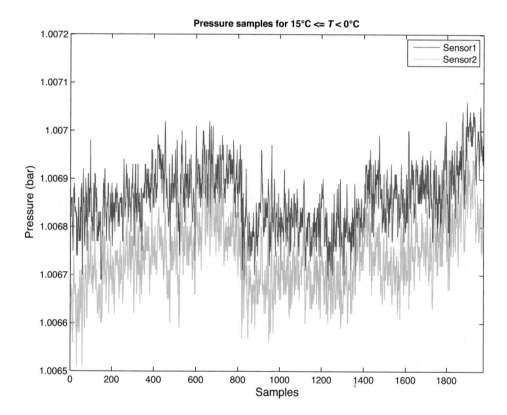

Figure 3.19: Pressure measurements for the temperature range 15–0°C.

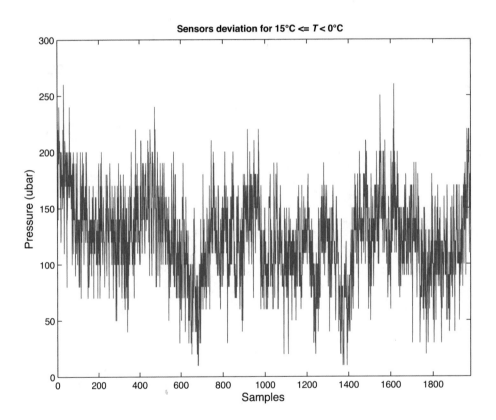

Figure 3.20: Deviation for the temperature range 15–0°C.

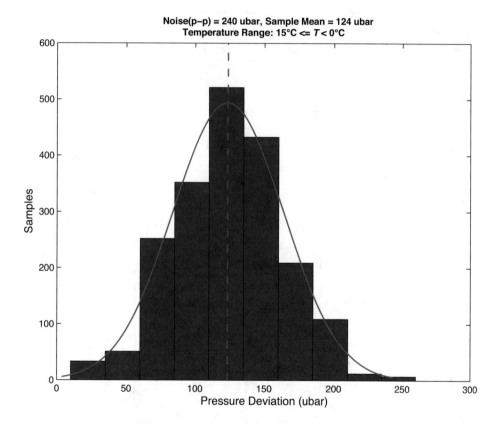

Figure 3.21: Histogram of pressure deviation for the temperature range 15–0°C.

It is worth mentioning that the number of samples is not fixed for each temperature range. This option would only be possible with the employment of an environmental chamber (of considerable cost), able to control the inspected range of temperature.

Sampling distribution (along with sample mean and p-p noise) of each temperature range is given in Figure 3.22. The selected color code is addressed to facilitate reading of the figure. In detail, colors red(R), green(G), blue(B), cyan(C), magenta(M), and yellow(Y), correspond to temperature ranges $R = 0$–15°C, $G = 15$–30°C, $B = 30$–40°C, $C = 40$–50°C, $M = 50$–60°C, and $Y = 0$–60°C (distribution of the latter range is not normalized, it is only depicted in this shape in order to give a hint about the total spread in deviation for the overall temperature range). This figure reveals that as temperature rises, deviation in sensors' output signal increases for this particular pair, as well as model of sensors.

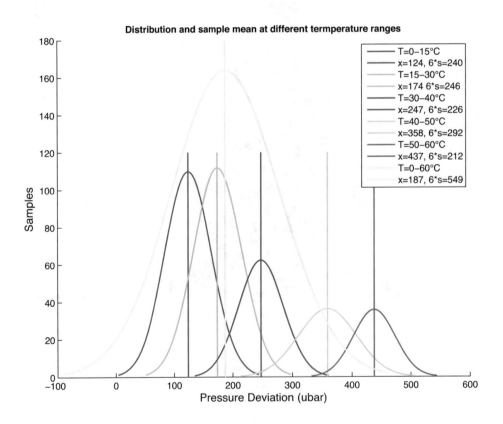

Figure 3.22: Temperature effect on the accuracy of estimator (sampling distribution).

This conclusion is made clearer in the bar graph depicted in Figure 3.23, where error bars (depicted in red line) illustrate the *standard error of the mean* (SEM). A more descriptive way

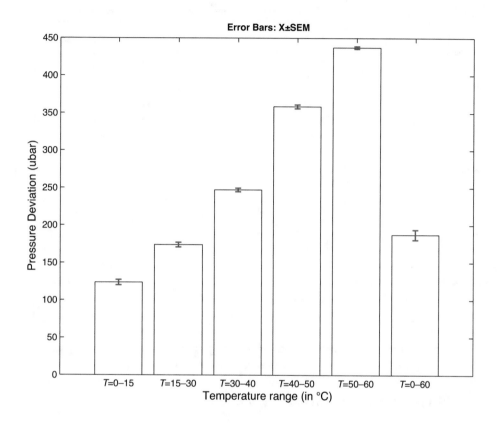

Figure 3.23: Temperature effect on the accuracy of estimator (error bars).

to graphically observe temperature influence on measurement accuracy is given by the box plot diagrams of Figure 3.24. Each box plot characterizes the sample data of each particular temperature range. The box plot splits into *quartiles* (Q), where the so-called *box* covers the central 50% of data among the lower and upper quartile (that is, Q1 and Q3), aka *interquartile range* (IQR). In reference to Figure 3.25, the size of IQR reveals the *precision* of measurement (i.e., smaller size is identical to better precision). The vertical line inside of the box is the *median* value of the data set, which illustrates the possible left/right *skewness* pattern of the dataset (i.e., fewer observations on the left/write area of the sample distribution). The turkey-style whiskers of box plot extend to a maximum 1.5xIQR beyond the box. The multiplier corresponds to approximately $\pm 2.7\sigma$, thereby incorporating the 99.3% of the data of a normalized distribution. Plotting points that are traced too far from the central value (i.e., beyond whiskers) is also possible, while those points are regularly referred to as *outliers*, that is, the red cross-marks of Figure 3.24. The box

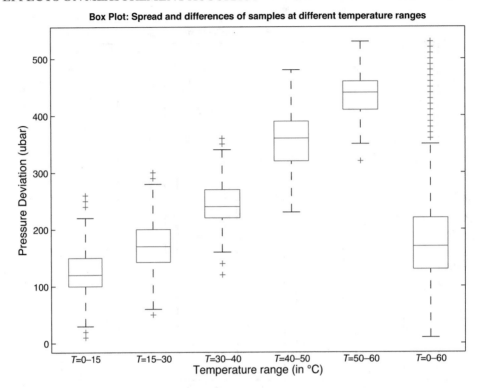

Figure 3.24: Temperature effect on the accuracy of estimator (box plots).

plot of the overall temperature range (i.e., 0–60°C) illustrates a significant number of samples around the pressure deviation 350–450 ubar.

In conclusion, while a system applying differential altimetry is less vulnerable to the possible variations of atmospheric pressure, the temperature can considerably affect measurement accuracy of the system. However, the addressing of a low-cost experiment can be proved an effective solution toward the proper design of a system employing MEMS barometers, able to sustain long-term measurement stability. In addition, many systems are designed to work in a specific range of temperature and if this range is identified and evaluated, it could be used to further increase measurement accuracy of the system. It should be noted, though, that this particular experimentation technique should be reiterated among different pairs of devices, as each MEMS barometric sensor is characterized by a unique pressure-response curve. In addition, the utilization of identical sensor devices in systems applying differential altimetry is recommended, as previous research has proven that sensors of diverse vendors respond in a different way to the effects of temperature [6].

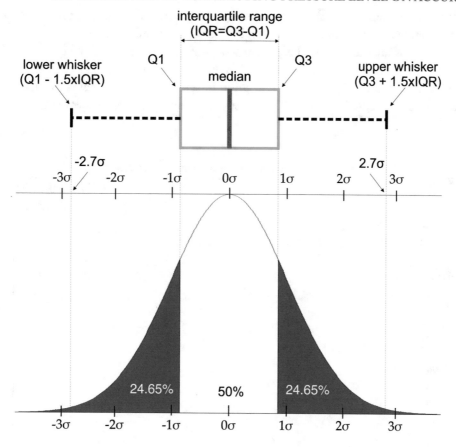

Figure 3.25: Box-and-whisker plot.

3.3   INFLUENCES OF THE EXISTING PRESSURE LEVEL ON ACCURACY

As mentioned earlier in the chapter, deviation in sensors' output signal (in a system applying differential altimetry measurement) is also affected by the existing acquired level of pressure. The following experiment reveals that (contrary to the influence of temperature) measurement accuracy features a random effect in consideration of the acquired pressure level (and for sensor devices BME280). To evaluate this effect on measurement accuracy we run the following experiment:

Inside of the airtight enclosure, which was presented in the previous chapter, we place the two wireless barometer devices. Using the vacuum handheld pump at the top of the enclosure, we extract the air till the pressure level falls approximately at 0.960 ubar (the latter value is determined by the measurements of BME280 sensors). We acquire 125 samples at room temperature and, thereafter, we increase the pressure level at 0.965 ubar and repeat the measurement process. Increment of the pressure level is performed through the appliance of a gentle mechanical force to the relief valve of the enclosure, which allows a quantity of air to be inserted in the enclosure (as depicted by Figure 3.26). Deviation in sensors' output signal is calculated for the pressure levels 0.960, 0.965, 0.970, 0.975, 0.980, 0.985 ubar, as depicted by Figures 3.27–3.32, respectively.

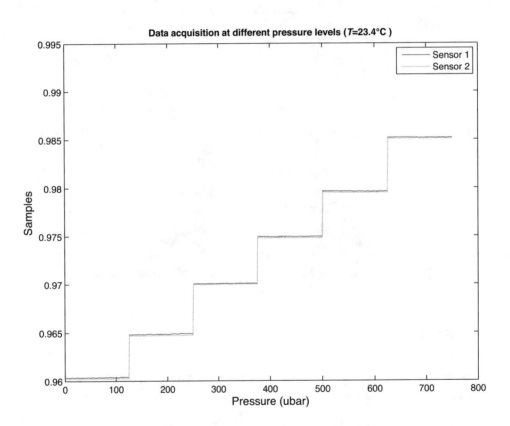

Figure 3.26: Data acquisition process.

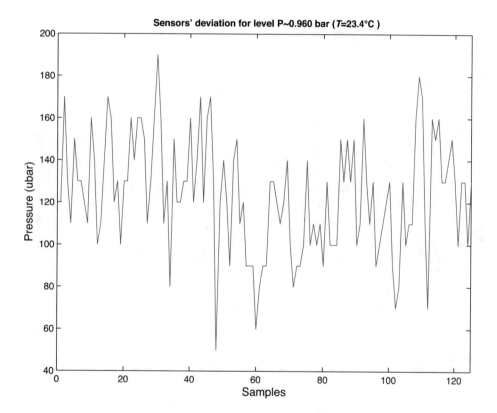

Figure 3.27: Deviation for the level of pressure 0.960 bar.

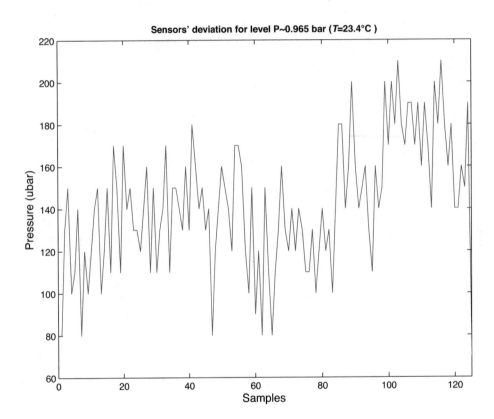

Figure 3.28: Deviation for the level of pressure 0.965 bar.

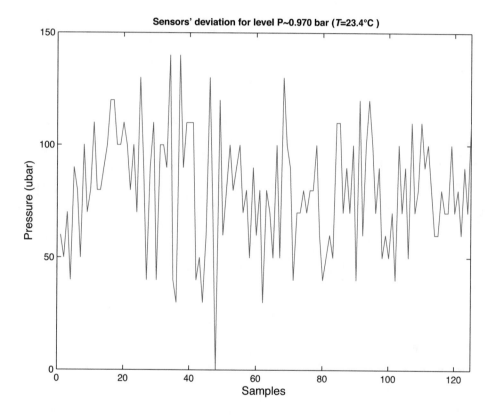

Figure 3.29: Deviation for the level of pressure 0.970 bar.

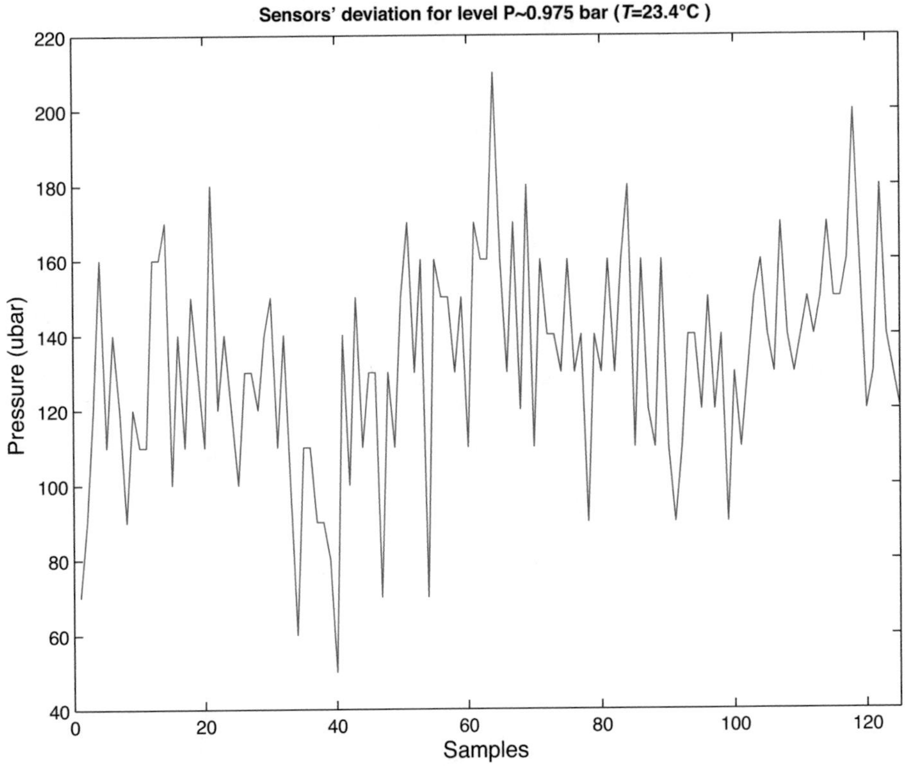

Figure 3.30: Deviation for the level of pressure 0.975 bar.

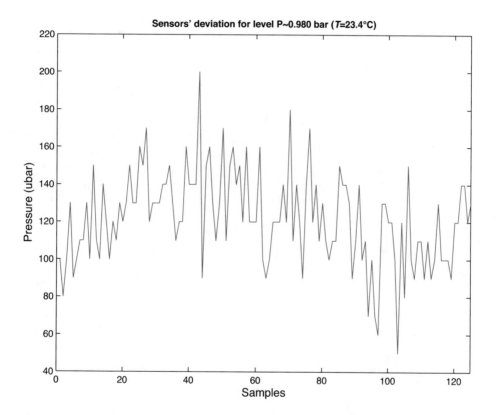

Figure 3.31: Deviation for the level of pressure 0.980 bar.

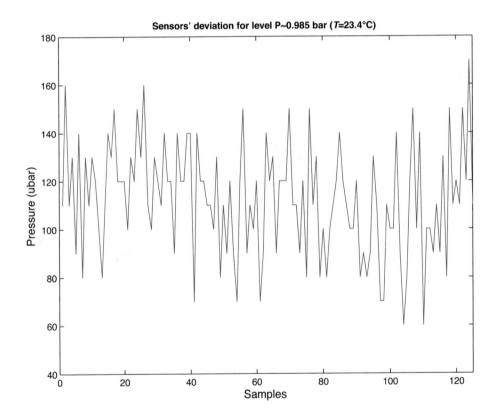

Figure 3.32: Deviation for the level of pressure 0.985 bar.

Figure 3.33 depicts the deviation in sensors' output signal for the overall set of pressure levels. The histogram of Figure 3.34 illustrates a normal distribution of the deviation signal. Sampling distribution (along with sample mean and p-p noise) of each pressure level is depicted in Figure 3.35, using the same color code as before, i.e., R, G, B, C, M, Y plus key (K) for the distribution of the overall dataset. Randomness in the influence of the sensing pressure is clearly illustrated by the random displacement of the box plot diagrams of Figure 3.36.

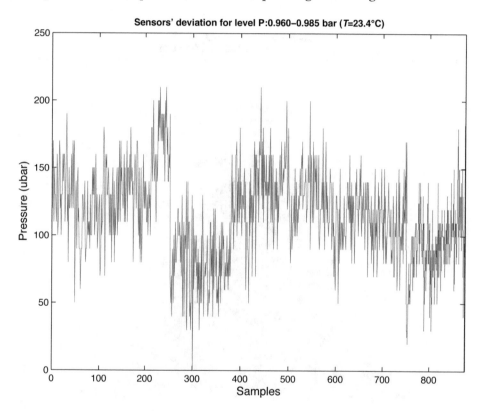

Figure 3.33: Deviation for all levels of pressure (i.e., 0.960–0.985 bar).

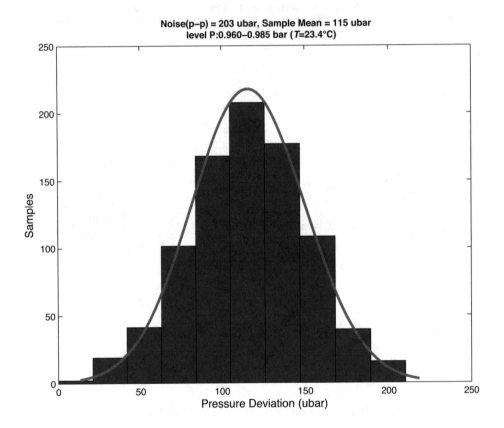

Figure 3.34: Histogram of pressure deviation for all levels (i.e., 0.960–0.985 bar).

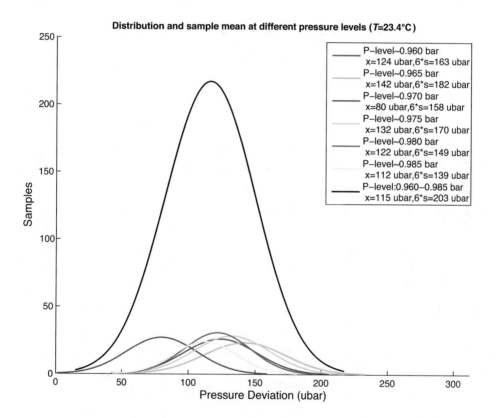

Figure 3.35: Effect of the existing pressure level on the accuracy of estimator (sampling distribution).

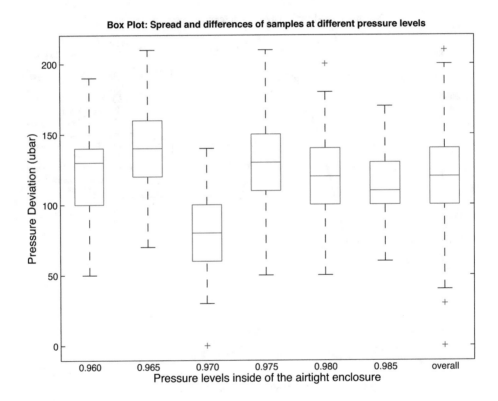

Figure 3.36: Effect of the existing pressure level on the accuracy of estimator (box plots).

REFERENCES

[1] *BME280 Combined Humidity and Pressure Sensor*, Bosch Sensortec, Germany, 2015.

[2] L. Binghao, B. Harvey, and T. Gallagher, Using barometers to determine the height for indoor positioning, in *Proc. of the International Conference on Indoor Positioning and Indoor Navigation*, Montbeliard-Belfort, France, pp. 1–7, 2013. DOI: 10.1109/ipin.2013.6817923.

[3] C. Bollmeyer, M. Pelka, H. Gehringy, and H. Hellbruck, Evaluation of radio based, optical and barometric localization for indoor altitude estimation in medical applications, in *Proc. of the International Conference on Indoor Positioning and Indoor Navigation*, Bexco, Busan, Korea, pp. 1–10, 2014. DOI: 10.1109/ipin.2014.7275477.

[4] W. Tang and Y-H Tsai, Barometric altimeter short-term accuracy analysis, *IEEE Aerospace and Electronic Systems Magazine*, 20(12), pp. 24–26, 2005. DOI: 10.1109/maes.2005.1576100.

[5] D. E. Bolanakis, K. T. Kotsis, and T. Laopoulos, A prototype wireless sensor network system for a comparative evaluation of differential and absolute barometric altimetry, *IEEE Aerospace and Electronic Systems Magazine*, 30(11), pp. 20–28, 2015. DOI: 10.1109/maes.2015.150013.

[6] D. E. Bolanakis, Evaluating performance of MEMS barometric pressure sensors in differential altimetry systems, *IEEE Aerospace and Electronic Systems Magazine*, manuscript sent to production.

CHAPTER 4

Height Acquisition and Measurement Analysis

This chapter applies vertical (absolute) distance determination with the employment of the proposed WSN system described earlier in the book. Measurement along with *Student's t-test* analysis toward the determination of the estimator's statistical significance is illustrated by this chapter.

4.1 MEASUREMENT ANALYSIS IN DIFFERENTIAL ALTIMETRY

The examples presented in this chapter explore accuracy in height determination in differential as well as single-ended altimetry. The measurement setup applying to both methods of altimetry is depicted in Figure 4.1.

In differential altimetry the two barometer devices are initially placed at reference position (Figure 4.1a) and measurements are obtained in order to determine existing deviation in sensors' output signal. Thereafter, the rover altimeter is moved at vertical (absolute) distance 88.5 cm lower than the based altimeter and the measurement process is repeated (Figures 4.1b and 4.1c). Pressure measurements are converted into altitude data using both the international barometric formula and hypsometric equation. Finally, the difference between the two altitude values is determined by the difference of the latter values, while applying a correction to the deviation signal identified by the former measurement process. On the other hand, in single-ended altimetry, only the measurements obtained from the rover (in the system) altimeter, are analyzed. Thereby, a comparative evaluation of the accuracy in measurements, derived by differential as well as single-ended altimetry, is possible.

Figure 4.2 presents the data acquisition process in LabVIEW-based GUI. The white graph depicts pressure measurements acquired by the base altimeter, while the red graph depicts data of the rover altimeter. At the half period of the overall data acquisition procedure an increment in the pressure graph of the rover altimeter is observed, as the latter device was (at that moment) moved at absolute height 0.885 m lower than the reference position (Figure 4.1c).[1] The figure examples given hereafter illustrate the measurement analysis procedure and, thereafter, representative samples are used to estimate the population statistics.

---

[1]As a reminder, air pressure decreases exponentially as elevation increases, and vice versa.

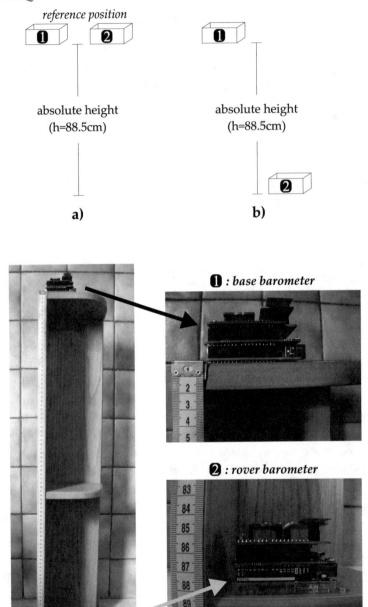

Figure 4.1: Vertical (absolute) distance measurement setup.

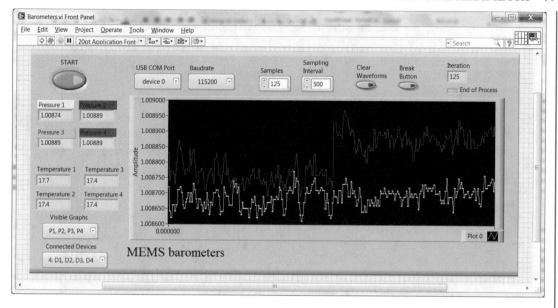

Figure 4.2: Data acquisition in LabVIEW-based GUI.

Figures 4.3 and 4.4 present the pressure measurements obtained from the two altimeters at reference position and height difference, respectively. The blue line depicts the atmospheric pressure data obtained from the rover altimeter, while the dotted red line depicts data acquired by the base altimeter. The deviation in sensors' output signal is determined by the computation of the absolute difference between these two graphs at reference position (Figure 4.3). Thereafter, corrections are applied to the pressure data of the base altimeter, which are depicted in the solid red line (for both figure examples).

Corrections are in line with the addition or subtraction of the average value of raised deviation to each particular pressure measurement acquired by the base altimeter. In this particular example, the pressure graph of the base altimeter (dotted red line) is reduced compared to the corresponding graph of rover altimeter (solid blue line) at reference position (Figure 4.3). Therefore, the average value of deviation is added to the former graph. That is, a boost relative to the existing average deviation in sensors' output signal is given to pressure graphs of base altimeter (for both examples of Figures 4.3 and 4.4). Hence, the pressure graphs depicted in solid red and solid blue lines of Figure 4.3 overlay each other, as they were expected to do since barometers acquired measurements at the exact same elevation. On the other hand, the solid red line of Figure 4.4 is reduced compared to the solid blue line, as the rover altimeter (which is in agreement with the latter graph) is positioned at altitude 0.885 m lower than the base altimeter.

The absolute height difference among the two sensors is determined by the conversion of pressure measurements (i.e., data depicted by the solid line graphs of Figure 4.4) into altitudes,

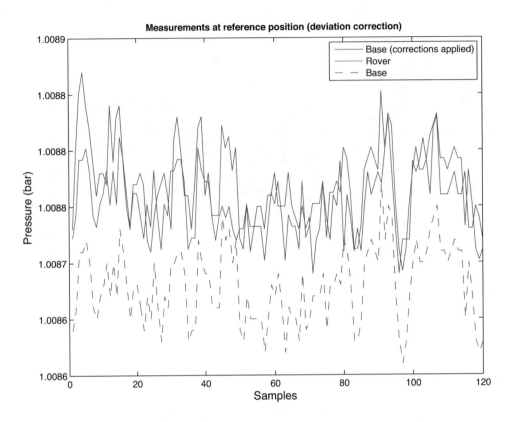

Figure 4.3: Pressure data of barometers at reference position (differential altimetry).

and thereafter the subtraction of lower from the higher altitude value acquired by the rover and base altimeter, respectively.

Figures 4.5 and 4.6 illustrate the absolute height graph between the two sensor devices, determined by international barometric formula and hypsometric equation, respectively. The identical shape of these two examples reveals that height graph generated from same set of measurements. However, the calculations generated by two different formulas produce a slightly different result in the average value of absolute vertical difference, which is given in the title of figures. The histogram of the height graph in consideration of the international barometric formula is depicted in Figures 4.7, where peak-to-peak noise (given in the figure title) is determined for the 99.7% of the overall dataset (that is, for six standard deviations).

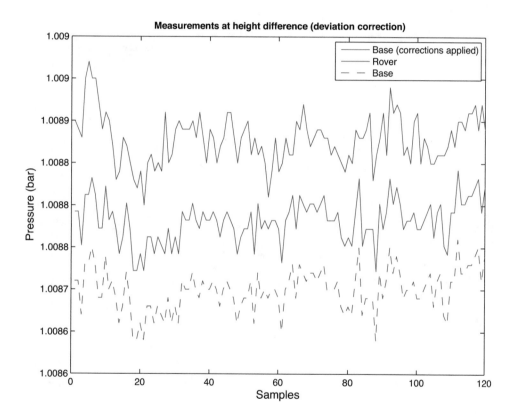

Figure 4.4: Pressure data of barometers at height difference (differential altimetry).

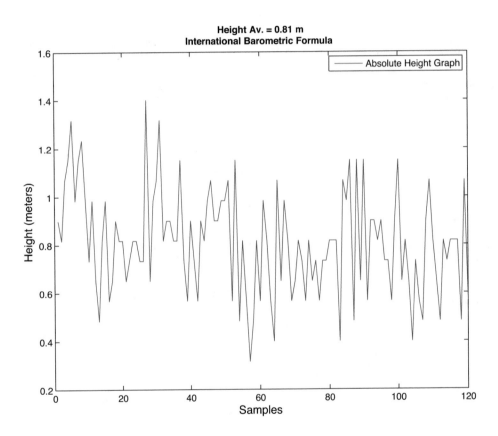

Figure 4.5: Height graph with international barometric formula (differential altimetry).

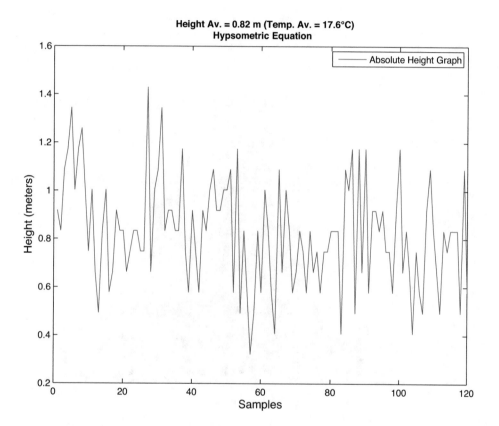

Figure 4.6: Height graph with hypsometric equation (differential altimetry).

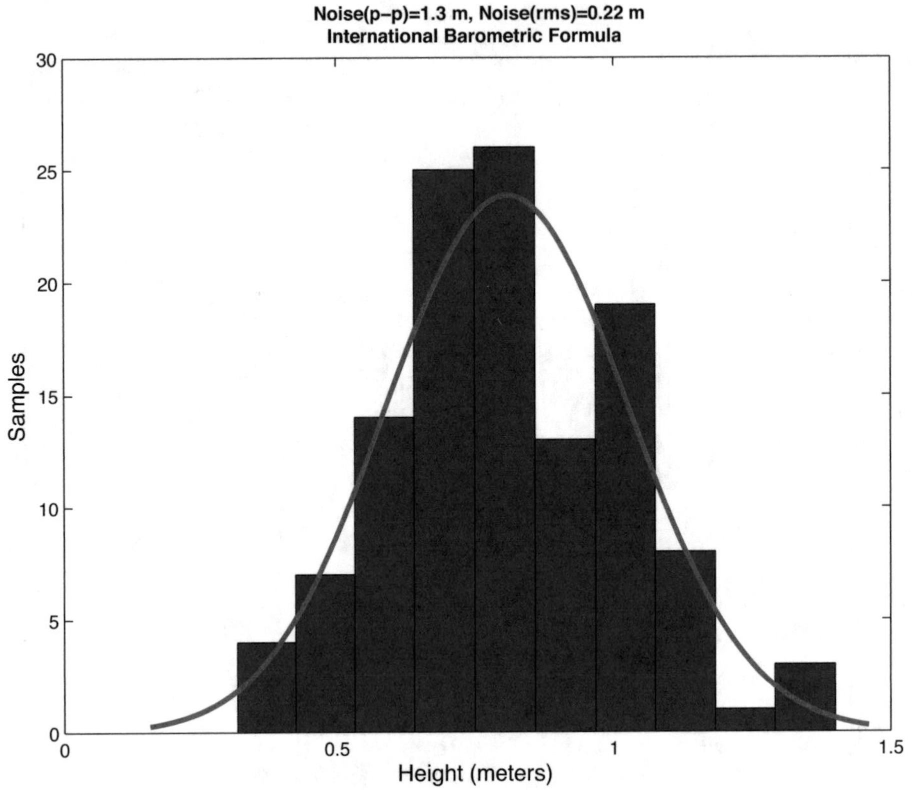

Figure 4.7: Histogram of height graph (differential altimetry).

4.2 MEASUREMENT ANALYSIS IN SINGLE-ENDED ALTIMETRY

The determination of absolute vertical difference among two positions constitutes a straight-forward procedure in single-ended altimetry. Figure 4.8 depicts measurements obtained from the rover altimeter, using the same dataset obtained before. The blue line depicts pressure data acquired at reference position, while the red line illustrates measurements at absolute altitude 0.885 m lower than the reference position.

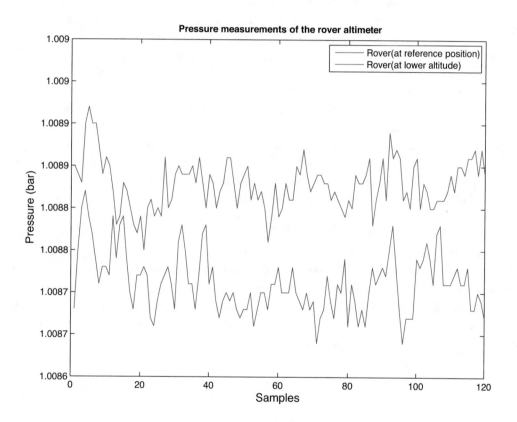

Figure 4.8: Pressure data of rover barometer (single-ended altimetry).

While no corrections are required for the determination of absolute height in single-ended altimetry, the vulnerability of this method to the atmospheric pressure variations can be intu-itively observed by the corresponding graphs of Figures 4.8. In detail, a decrement in the ambient air pressure is observed (i.e., from 1.0088 down to 1.0087 bar which corresponds to a drop of 100 ubar), during the data acquisition of the rover barometer at reference position, depicted by the graph in blue color. On the other hand, when the rover barometer acquires data at lower—by 0.885 m from the reference position—altitude, no variations in barometric pressure are present (i.e., pressure remains stable at approximately 1.0089 bar) as depicted by the graph in red color. For that reason, the difference between the two graphs is greater for the samples in the second half period (i.e., samples 61–120) than it is for the first half period.

The influence of pressure variations in single-ended altimetry can be observed from the height graph given in Figure 4.9. This plot is generated by the conversion of pressure graphs in altitudes with the employment of the international barometric formula and thereafter, the lower altitude is subtracted from the higher altitude.[2] The increment in the average value of height graph in the second half period (i.e., samples 61–120) generates an overall average value of absolute height equal to 0.97 m. As a reminder, the corresponding measurement analysis in differential altimetry generated an average value of absolute height equal to 0.81 m. Given that the theoretical value of height corresponds to 0.885 m, the percent error formula produces 8.47% height error in differential altimetry and 9.6% error in single-ended altimetry.

A small difference in height error is observed for the two methods of altimetry. However, the vulnerability of single-ended altimetry to the atmospheric pressure variations may affect absolute height measurements in the sense of short-term stability, as well. This feature of single-ended altimetry can be further verified by the left-skewed distribution (identified by the long left tail) of Figure 4.10. The histogram is generated by the height graph of Figure 4.9, which would be a normal distribution if no variation were present in the ambient air pressure acquisition.

Inferentially, it could be said that while the deviation in the two sensors' output signal can affect long-term measurement stability in differential altimetry, there are particular tests (described in the previous chapter) which can be addressed so as to sustain long-term stability conditions. On the other hand, the effects of the truly random variations in the ambient air pressure allow no further predictions to be made in order to assure long-term stability of single-ended altimetry. Moreover, this is the main reason this method is considered inappropriate for vertical distance measurements in outdoor environments, even for a short-term measurement period, because of the excess variations in atmospheric pressure [1].

[2]For a detailed description of the absolute height determination you may refer to Formula 1.1.

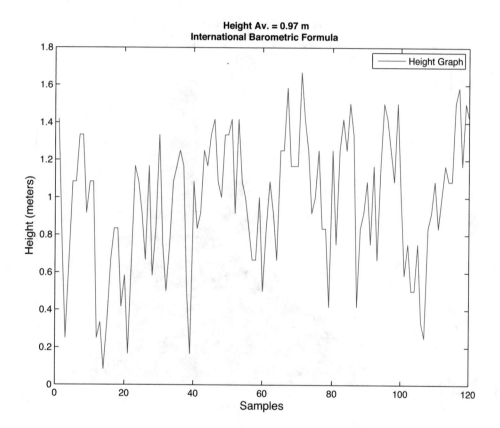

Figure 4.9: Height graph (single-ended altimetry).

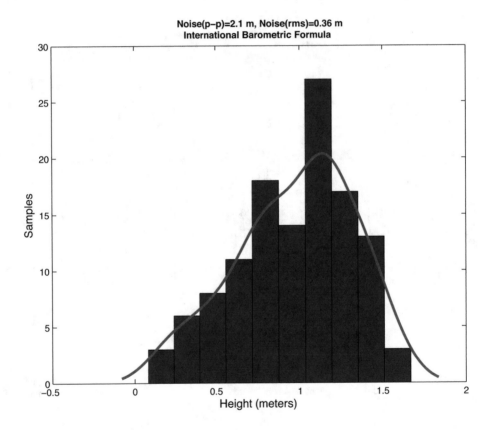

Figure 4.10: Histogram of height graph (single-ended altimetry).

4.3   STATISTICAL SIGNIFICANCE OF THE ESTIMATOR

Hereafter we present the results of repeated measurements, using the aforementioned setup for the acquisition of 0.885 m absolute height difference. Then the appropriate statistical analysis attempts to quantitatively model the role of chance in measurements, and associate uncertainty to the outcome of experiments (as the reproduced experiment almost never obtains identical results) [2].

We have discussed earlier in this book that MEMS barometric sensors obtain a Gaussian white noise in their output. When acquiring absolute height difference with MEMS barometers we are familiar with the population mean (where, $\mu = 0.885$ m for this particular experiment), yet the standard deviation of population (σ) is unknown. Because we are sampling from a population assumed to be normally distributed and whose standard deviation is unknown, the following statistical tests (addressed to determine statistical significance of the acquired samples) apply to *Student's t-test* and, in particular, to *one sample (two-tail) t-test*.

The absolute height of 0.885 m was acquired 10 successive times (of 120 samples for each particular set) using the procedure describe earlier in this chapter. Figures 4.11 and 4.12 present the overall measurements when barometers were placed at reference position and height difference, respectively. The low level of atmospheric pressure ($\cong 0.986$ bar) is due to the fact that measurements were obtained on a rainy day.

Tables 4.1 and 4.2 summarize the results obtained from the differential altimetry method, when performing calculations with the international barometric formula and hypsometric equation, respectively. Columns 1 and 2 provide information about the population, while the consecutive number of *3rd* column represents the corresponding measurement sample. The natural estimators of population mean (aka *sample mean*) and population standard deviation (aka *sample standard deviation*) are given in columns 4 and 5, respectively. Absolute height error of each sample, determined by the percent error formula (i.e., Formula 4.1), is illustrated in column 6. Finally, the average values of the 10 samples, that is, mean, deviation, and %error, are, respectively, presented in columns 7–9. It is worth noting that negligible differences are observed in calculations performed with the international barometric formula (Table 4.1) and hypsometric equation (Table 4.2), with the former producing slightly better performance.

$$\text{Percent Error} = \frac{|\text{Experimental - Theoretical}|}{\text{Theoretical}} \times 100$$

Formula 4.1: Percent error (%error).

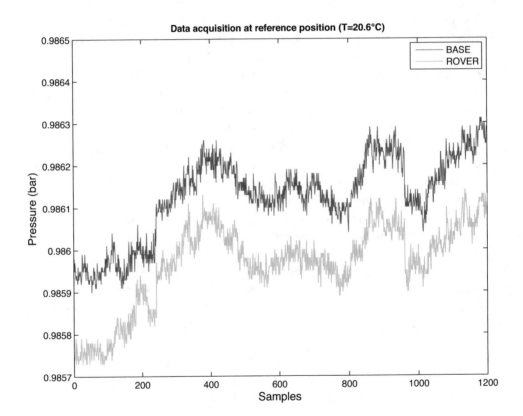

Figure 4.11: Ten measurement sets acquiring 0.885 m absolute height (reference position).

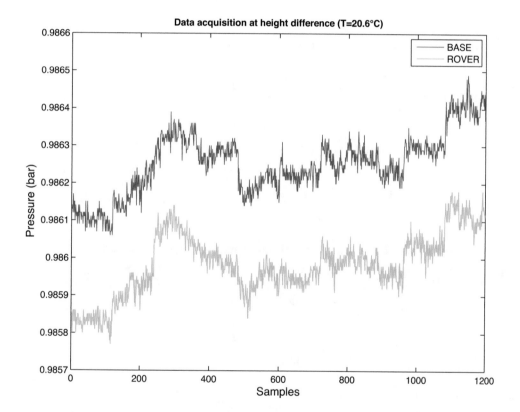

Figure 4.12: Ten measurement sets acquiring 0.885 m absolute height (height difference).

Table 4.1: Results obtained from differential altimetry using international barometric formula

μ	σ	n	$\bar{x}(n)$	s(n)	% Error (n)	$\bar{x}$	s	% Error
0.885 m	Unknown	1	0.68 m	0.22 m	23.2%	0.92 m	0.26 m	11.7%
		2	1.11 m	0.29 m	25%			
		3	0.88 m	0.33 m	0.9%			
		4	1.11 m	0.29 m	25.8%			
		5	0.76 m	0.25 m	14.5%			
		6	0.93 m	0.19 m	5.3%			
		7	0.98 m	0.23 m	11.1%			
		8	0.87 m	0.24 m	1.8%			
		9	0.93 m	0.23 m	4.9%			
		10	0.93 m	0.28 m	4.6%			

Table 4.2: Results obtained from differential altimetry using hypsometric equation

μ	σ	n	$\bar{x}(n)$	s(n)	% Error (n)	$\bar{x}$	s	% Error
0.885 m	Unknown	1	0.70 m	0.23 m	23.2%	0.95 m	0.27 m	13.7%
		2	1.14 m	0.30 m	25%			
		3	0.91 m	0.34 m	0.9%			
		4	1.15 m	0.30 m	25.8%			
		5	0.78 m	0.26 m	14.5%			
		6	0.97 m	0.20 m	5.3%			
		7	1.02 m	0.24 m	11.1%			
		8	0.90 m	0.25 m	1.8%			
		9	0.96 m	0.24 m	4.9%			
		10	0.96 m	0.30 m	4.6%			

The results obtained from single-ended altimetry, using the international barometric formula, are summarized in Table 4.3. On the other hand, Table 4.4 features the calculations of the hypsometric equation. Contrary to the previous case, single-ended altimetry reveals somewhat better performance when computations are in agreement with the hypsometric equation. It is worth mentioning that the empirical international barometric formula, which assumes a fixed temperature level of 15°C, is considered adequate for the (absolute) height determination [3]. However, if the sensed temperature is far from the reference value of 15°C, it is recommended using the hypsometric equation instead.

Table 4.3: Results obtained from single-ended altimetry using international barometric formula

μ	σ	n	$\bar{x}(n)$	$s(n)$	% Error (n)	$\bar{x}$	s	% Error
0.885 m	Unknown	1	0.61 m	0.23 m	30.7%	0.48 m	0.34 m	45.3%
		2	0.60 m	0.35 m	32%			
		3	0.72 m	0.49 m	18.6%			
		4	0.46 m	0.27 m	47.7%			
		5	0.20 m	0.39 m	77.5%			
		6	0.24 m	0.28 m	73.2%			
		7	0.34 m	0.30 m	61.3%			
		8	0.76 m	0.32 m	14.2%			
		9	0.48 m	0.36 m	45.3%			
		10	0.42 m	0.43 m	52.4%			

Table 4.4: Results obtained from single-ended altimetry using hypsometric equation

μ	σ	n	$\bar{x}(n)$	$s(n)$	% Error (n)	$\bar{x}$	s	% Error
0.885 m	Unknown	1	0.63 m	0.23 m	30.7%	0.50 m	0.35 m	43.3%
		2	0.62 m	0.36 m	32%			
		3	0.75 m	0.51 m	18.6%			
		4	0.48 m	0.28 m	47.7%			
		5	0.21 m	0.40 m	77.5%			
		6	0.25 m	0.29 m	73.2%			
		7	0.36 m	0.31 m	61.3%			
		8	0.79 m	0.33 m	14.2%			
		9	0.50 m	0.37 m	45.3%			
		10	0.44 m	0.44 m	52.4%			

From the above tables it is quite obvious that differential barometric altimetry is of improved accuracy compared to single-ended altimetry. Nevertheless, the following one-sample t-test on these two methods of altimetry allows us to explore whether the sample data could come from a normal distribution with given $\mu = 0.885$ and of unknown σ [4].

Formula 4.2a,b states the null and alternative hypothesis, respectively, as we explore if some difference exists between the sample mean and the expected value of 0.885 m. This hypothesis will be answered with an *alpha level* (aka *tail probability*) equal to 0.05 ($a = 0.05$), in a two tail test (Figures 4.13). Thereby, we are going to construct the middle 95% range where sample mean is expected to be found, aka *confidence interval* (CI). If the observed mean is outside of that area, we will conclude that it is different from the population mean and hence, we are going to *reject the null hypothesis*. Otherwise, we state that we *fail to reject the null hypothesis* (which would be the desired outcome).

$$a)\ H_0 : m = \mu$$
$$b)\ H_1 : m \neq \mu$$
$$where,\ \mu = 0.885\ m$$

Formula 4.2: Null (a) and alternative (b) hypotheses.

The 95% CI is determined by Formula 4.3, where the *standard error of the mean (SEM)* denotes (in consideration of the acquired number of samples) how precisely the sample mean estimates population mean. Lower values of SEM indicate more precise estimates, and vice versa.

$$\bar{x} \pm \left(t_{a/2} \right) \cdot \frac{s}{\sqrt{n}}\ ,\ where$$

s : standard deviation of the sample mean $(\bar{x})$

$\dfrac{s}{\sqrt{n}}$: standard error of the mean (SEM)

Formula 4.3: Constructing confidence interval about the sample mean $(\bar{x})$.

The t distribution critical value is obtained from t table (i.e., Table 4.5). The t critical value is determined by the *degrees of freedom* (df) indicating the corresponding row of t table, as well as tail probability indicating the column of table. Columns of Table 4.5 hold values of one-tail probability, which in our example is equal to 0.025. Degrees of freedom in our examples is equal to $n - 1 = 9$ (where n, the number of samples of each set).

Having determined CI, the t value of each measurement set is determined by Formula 4.4. The generated result should fall within the range: $-2.262 < t < +2.262$. If the latter condition is verified, then we fail to reject the null hypothesis. The calculated t value together with the employment of t-table (Table 4.5) could be addressed for the identification of the probability

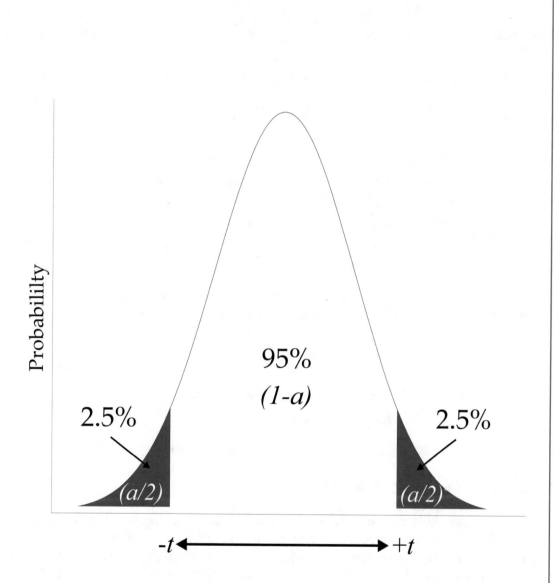

Figure 4.13: Two tail t-test, where $a = 0.05$ (decision rule of the current tests).

Table 4.5: Table t (critical values of t distribution)

df	Tail Probability (a)						
	0.10	0.05	0.025	0.01	0.005	0.001	0.0005
1	3.078	6.314	12.71	31.82	63.66	318.3	636.6
2	1.886	2.920	4.303	6.965	9.925	22.32	31.60
3	1.638	2.353	3.182	4.451	5.841	10.22	12.92
4	1.533	2.132	2.776	3.747	4.604	7.173	8.610
5	1.476	2.015	2.571	3.365	4.032	5.893	6.869
6	1.440	1.943	2.447	3.143	3.707	5.208	5.959
7	1.415	1.895	2.365	2.998	3.499	4.785	5.408
8	1.397	1.860	2.306	2.896	3.355	4.501	5.041
9	1.383	1.833	2.262	2.821	3.250	4.297	4.781
10	1.372	1.812	2.228	2.764	3.169	4.144	4.587

range, as well. If the identified p value is greater than the alpha level (where $a = 0.05$ for this particular example), we conclude that we fail to reject the null hypothesis.

$$t = \frac{x - \mu}{\frac{s}{\sqrt{n}}} \quad , \text{where}$$

s : standard deviation of the sample mean $(\bar{x})$

$\frac{s}{\sqrt{n}}$: standard error of the mean (SEM)

μ : population mean; n : number of samples

Formula 4.4: Determining t critical value.

From the previous equations it is quite obvious that t-tests require lots of computation effort. However, the test can be rapidly arranged when using an appropriate software program (e.g., Excel, Matlab, etc.). For instance, Matlab code line *[h,p,ci,stats] = ttest(x,u)*, where x is the array incorporating the sample means and u the value of population mean, returns the following data:

- h: fail to reject ($h = 0$) *or* reject null hypothesis ($h = 1$) at the 5% significant level

- p: value of probability (in the range $[0, 1]$)

- ci: confidence interval

- *tstat*: value of the *t* statistic

- *df*: degrees of freedom of the test

- *sd*: estimated population standard deviation

The one sample *t*-test, summarized in Table 4.6, illustrates that there is a statistically significant difference between the sample mean and population mean values in single-ended altimetry. According to the theory covered previously in the book, the latter conclusion is in agreement with the vulnerability of single-ended altimetry method to pressure variations. Accuracy can be severely disturbed even for short-term measurements, like in this particular example where each set lasted for about 10–15 min. On the other hand, the differential altimetry method is more immune to such effects, while the proper design of a system (according to the illustrative examples covered earlier in the book) could sustain long-term measurement stability, as well.

Table 4.6: One sample *t*-test

Method	Height Determination with:	
	International Barometric Formula	Hypsometric Equation
Differential Altimetry (df = 9)	h = 0	h = 0
	p = 0.4730	p = 0.1775
	ci(m) = 0.8202, 1.0139	ci(m) = 0.8496, 1.0502
	t = 0.749	t = 1.4631
	sd(m) = 0.1354	sd(m) = 0.1403
Single-ended Altimetry (df = 9)	h = 1	h = 1
	p = 95.697×10^{-6}	p = 172.54×10^{-6}
	ci(m) = 0.3475, 0.6209	ci(m) = 0.3599, 0.6430
	t = -6.6320	t = -6.1315
	sd(m) = 0.1911	sd(m) = 0.1978

REFERENCES

[1] D. E. Bolanakis, K. T. Kotsis, and T. Laopoulos, A prototype wireless sensor network system for a comparative evaluation of differential and absolute barometric altimetry, *IEEE Aerospace and Electronic Systems Magazine*, 30(11), pp. 20–28, 2015. DOI: 10.1109/maes.2015.150013.

[2] M. Krzywinski and N. Altmanand, Importance of being uncertain, *Nature Methods*, 10(9), pp. 809–810, 2013. DOI: doi:10.1038/nmeth.2613.

[3] D. E. Bolanakis, K. T. Kotsis, and T. Laopoulos, Temperature influence on differential barometric altitude measurements, in *Proc. of the 8th IEEE International Conference on Intelligent Data Acquisition and Advanced Computing Systems: Technology and Applications (IDAACS'2015)*, Warsaw, Poland, pp. 120–124, 2015. DOI: 10.1109/idaacs.2015.7340711.

[4] M. Krzywinski and N. Altmanand, Significance, *p* values and *t*-tests, *Nature Methods*, 10(11), pp. 1041–1042, 2013. DOI: 10.1038/nmeth.2698.

APPENDIX A

Matlab Script Code

This appendix provides information on the Matlab script code, which can be used as a reference guide toward the measurement analysis of methods addressed by the book.

A.1 MEASUREMENT ANALYSIS OF TEMPERATURE INFLUENCE (MATLAB CODE)

The following script code can be used as a reference guide for analyzing the temperature influence on measurement accuracy in differential altimetry (as explored in Chapter 3). The measurement files analyzed in Matlab consist of four columns; the former two incorporate pressure and temperature data of barometer 1, while the latter two encompass the corresponding data of sensor device 2. Hereafter we present a sample of the first five measurements of the current file under analysis (consisting of 5,350 samples).

1.00640	60.0	1.00595	60.7
1.00645	60.0	1.00600	60.7
1.00645	60.0	1.00600	60.7
1.00647	59.9	1.00598	60.7
1.00647	59.9	1.00601	60.7

To run the script code drag and drop the measurement file, which should be named "Pressure.txt," into Matlab Workspace. Then copy the script code into Matlab Command Window. The following code lines generate three figures for each one of the six predefined temperature ranges (that is, 0–15°C, 15–30°C, 30–40°C, 40–50°C, 50–60°C, and the overall 0–60°C). Thereby, 18 total figures are created in the default working folder of Matlab, illustrating:

1. The pressure samples obtained from the two barometer devices (Fig.x-1__Pressure-samples.tif);

2. Deviation in sensors' output signal in consideration of the acquired pressure samples (Fig.x-2__Sensors-deviation.tif);

3. The histogram of deviation in sensors' output signal (Fig.x-3__Histogram-of-pressure-deviation.tif); where the number x in the filename of all three figures depends on the current temperature range (i.e., admitting numbers from 1–6 in this example).

A brief description of these code lines is as follows. Code line 1 defines the number of divided intervals (aka *bins*) of the overall range of values to be used in the histogram plots. Lines 2–8 define the key values of temperature of the explored temperature ranges. Line 9 identifies the depth of measurement file (based on the number of rows found in the first column). The *for-loop* statement of code lines 10–36 separates the file measurements into arrays relative to the employed temperature range. Hence, array1 obtains pressure (column 1) and temperature (column 2) data of sensor device 1, as well as pressure data of sensor device 2 (column 3), for the temperature range 0.1–15°C (code lines 11–15). Based on this scheme, array2 (lines 16–20), array3 (lines 21–25), array4 (lines 26–30), and array5 (lines 31–35) incorporate measurements for the temperature ranges 15.1–30°C, 30.1–40°C, 40.1–50°C and 50.1–60°C, respectively, (with the higher temperature value traced at the top of each array). The generated arrays are of same depth but because each temperature may employ a different number of samples, code lines 37–41 remove the possible empty rows from each array. At this point it is worth mentioning that the temperature range is determined by the acquired temperature obtained from barometer 1 (found in column 2).

```
1       histBins=10;
2       TempRanges=5;
3       T1=0;
4       T2=15;
5       T3=30;
6       T4=40;
7       T5=50;
8       T6=60;

9       Samples = numel(Pressure(:,1));
10      for i=1:Samples
11        if ( (Pressure(i,2)>T1) && (Pressure(i,2)<=T2) )
12        array1(i,1)=Pressure(i,1);
13        array1(i,2)=Pressure(i,2);
14        array1(i,3)=Pressure(i,3);
15        end
16        if ( (Pressure(i,2)>T2) && (Pressure(i,2)<=T3) )
17        array2(i,1)=Pressure(i,1);
18        array2(i,2)=Pressure(i,2);
19        array2(i,3)=Pressure(i,3);
20        end
21        if ( (Pressure(i,2)>T3) && (Pressure(i,2)<=T4) )
22        array3(i,1)=Pressure(i,1);
23        array3(i,2)=Pressure(i,2);
```

```
24        array3(i,3)=Pressure(i,3);
25        end
26        if ( (Pressure(i,2)>T4) && (Pressure(i,2)<=T5) )
27        array4(i,1)=Pressure(i,1);
28        array4(i,2)=Pressure(i,2);
29        array4(i,3)=Pressure(i,3);
30        end
31        if ( (Pressure(i,2)>T5) && (Pressure(i,2)<=T6) )
32        array5(i,1)=Pressure(i,1);
33        array5(i,2)=Pressure(i,2);
34        array5(i,3)=Pressure(i,3);
35        end
36      end

37      array1( ~any(array1,2), : ) = [];
38      array2( ~any(array2,2), : ) = [];
39      array3( ~any(array3,2), : ) = [];
40      array4( ~any(array4,2), : ) = [];
41      array5( ~any(array5,2), : ) = [];


42      for i=1:(TempRanges+1)

43      switch i
44              case 1
45                      currentArray = array1;
46                      Tlow=T1;
47                      Thigh=T2;
48              case 2
49                      currentArray = array2;
50                      Tlow=T2;
51                      Thigh=T3;
52              case 3
53                      currentArray = array3;
54                      Tlow=T3;
55                      Thigh=T4;
56              case 4
57                      currentArray = array4;
58                      Tlow=T4;
```

```matlab
59                    Thigh=T5;
60            case 5
61                    currentArray = array5;
62                    Tlow=T5;
63                    Thigh=T6;
64            otherwise
65                    currentArray = Pressure;
66                    Tlow=T1;
67                    Thigh=T6;
68        end

69        currentArray_Samples = numel(currentArray(:,1));
70        P1=currentArray(:,1);
71        P2=currentArray(:,3);
72        Pdif_ubar=1000000*abs(P1-P2);
73        x=mean(Pdif_ubar);
74        s=std(Pdif_ubar);
75        SEM=s/sqrt(x);
76        xi(i)=x;
77        pp(i)=6*s;
78        SEMi(i)=SEM;

79        figure;
80        plot(P1,'r');
81        hold on;
82        plot(P2,'g');
83        xlim([0 currentArray_Samples]);
84        yl=ylim;
85        ylim([ yl(1) (yl(2)*1.0001) ]);
86        legend('Sensor1','Sensor2');
87        xlabel('Samples', 'FontSize', 12);
88        ylabel('Pressure (bar)', 'FontSize', 12);
89        title(sprintf('Pressure samples for %d %cC <= T < %d %cC',...
          Thigh, char(176), Tlow, char(176)), 'FontSize', 10, ...
          'FontWeight', 'bold');
90        eval(['print -dtiff Fig.' num2str(i) '-1__Pressure-samples'...
          '.tif' ]);

91        figure;
```

```
92     plot(Pdif_ubar,'b');
93     xlim([0 currentArray_Samples]);
94     xlabel('Samples', 'FontSize', 12);
95     ylabel('Pressure (ubar)', 'FontSize', 12);
96     title(sprintf('Sensors deviation for %d %cC <= T < %d %cC', Thigh,...
       char(176), Tlow, char(176)), 'FontSize', 10, 'FontWeight', 'bold');
97     eval(['print -dtiff Fig.' num2str(i) '-2__Sensors-deviation'...
       '.tif' ]);

98     figure;
99     histfit(Pdif_ubar,histBins);
100    xlabel('Pressure Deviation (ubar)', 'FontSize', 12);
101    ylabel('Samples', 'FontSize', 12);
102    ppNoise=6*s;
       title(sprintf('Noise(p-p) = %.0fubar, Sample Mean = %.0fubar\n...
       Temperature Range: %d %cC <= T < %d %cC', ppNoise, x, Thigh,...
       char(176), Tlow, char(176)), 'FontSize', 10, FontWeight', 'bold');
103    hLines = findobj('Type','Line');
104    set(hLines(1),'LineWidth',1);
105    hold on;
106    plot([x,x],ylim,'r--','LineWidth',1);
107    eval(['print -dtiff Fig.' num2str(i)...
       '-3__Histogram-of-pressure-deviation' '.tif' ]);

108    end
```

As soon as the arrays (holding data of each temperature range) are created, the figures are plotted within the *for-loop* statement of code lines 42–109 (addressed to run for each temperature range). At first, a *switch-case* statement configures the array and temperature (low and high) values at which the analysis is fetched (lines 43–68). Some computations for each particular array are performed in code lines 69–78 (where depth of each array is determined in code line 69). In detail, deviation in sensors' output signal is estimated in line 72, while average value (x) of the latter signal along with the standard deviation (s) and standard error of the mean (SEM) are, respectively, performed in lines 73–75. Code lines 76–78 collect x, peak-to-peak (pp) and SEM values of each temperature range into arrays of identical names (where conversion of rms to pp noise is determined for six standard deviations, i.e., 3s of either side of sample mean).

Pressure measurements (for each temperature range) are plotted in code lines 79–90. Pressure arisen from barometer 1 is depicted in red color (line 80), while pressure data of barometer 2 are depicted in green color (line 82). Code line 83 defines the limits of the x axis based on the depth of each array (calculated earlier in code line 69). Then the upper limit of the y axis is extended in line 85 (as defined by the multiplier 1.0001), so that the legend of the figure (line 86) does not overlap with pressure graphs. Labels of x,y axes as well as title of figure are generated in code lines 87–89. It is worth noting that *char(176)* corresponds to the symbol of Celsius degrees (°), defined by *Latin-1 Supplement* of *Unicode Characters* [1, 2]. Code line 90 prints the figure in compressed .tif format, while integrating into the filename a number identical to the loop counter. The three dots (…) used in code lines 89 and 90 are addressed to split large Matlab code lines. Deviation in sensors' output signal is plotted in code lines 91–97, while its histogram is plotted lines 98–107. In consideration of the latter figure, code lines 103–106 plot the mean value of sensors' deviation in lines 103–106. The next lines of script code generate five figures, described hereafter.

```
109   figure;
110   for i=1:(TempRanges+1)

111      switch i
112           case 1
113                currentArray = array1;
114                color='r';
115           case 2
116                currentArray = array2;
117                color='g';
118           case 3
119                currentArray = array3;
120                color='b';
121           case 4
122                currentArray = array4;
123                color='c';
124           case 5
125                currentArray = array5;
126                color='m';
127           otherwise
128                currentArray = Pressure;
129                color='y';
130      end
131   hold on;
132   h=histfit(1000000*abs(currentArray(:,1)-currentArray(:,3)));
133   delete(h(1));
134   hold on;
135   hLines = findobj('Type','Line');
136   set(hLines(1),'Color',color,'LineWidth',1);
```

```
137    x=mean(1000000*abs(currentArray(:,1)-currentArray(:,3)));
138    hold on;
139    plot([x,x],ylim,color,'LineWidth',1);

140    end

141    xl=xlim;
142    xlim([ xl(1) (xl(2)*1.001) ]);
143    yl=ylim;
144    ylim([ yl(1) (yl(2)*1.001) ]);
145    legend( sprintf('T=%d-%d %cC', T1,T2,char(176)),...
           sprintf('x=%.0f,6*s=%.0f', xi(1), pp(1) ),...
           sprintf('T=%d-%d %cC', T2,T3,char(176)),...
           sprintf('x=%.0f 6*s=%.0f', xi(2), pp(2) ),...
           sprintf('T=%d-%d %cC', T3,T4,char(176)),...
           sprintf('x=%.0f, 6*s=%.0f', xi(3), pp(3) ),...
           sprintf('T=%d-%d %cC', T4,T5,char(176)),...
           sprintf('x=%.0f, 6*s=%.0f', xi(4), pp(4) ),...
           sprintf('T=%d-%d %cC', T5,T6,char(176)),...
           sprintf('x=%.0f, 6*s=%.0f', xi(5), pp(5) ),...
           sprintf('T=%d-%d %cC', T1,T6,char(176)),...
           sprintf('x=%.0f, 6*s=%.0f', xi(6), pp(6) ) );
146    ylabel('Samples', 'FontSize', 12);
147    xlabel('Pressure Deviation (ubar)', 'FontSize', 12);
148    title(sprintf('Distribution and sample mean at different...
           termperature ranges'), 'FontSize', 10, 'FontWeight', 'bold');
149    print -dtiff Fig.A__Distribution-for-all-temperature-ranges.tif;
150    figure;
151    name = {'T=0-15';'T=15-30';'T=30-40';'T=40-50';'T=50-60';'T=0-60'};
152    x=[1:6];
153    bGraph1=bar(x,xi);
154    set(gca,'xticklabel',name);
155    set(bGraph1,'FaceColor',[1 1 1]);
156    hold on;
157    errorbar(1:TempRanges+1,xi,SEMi,'.r','LineWidth',1);
158    xlabel(sprintf('Temperature range (in %cC)', char(176)),...
           'FontSize', 12);
159    ylabel('Pressure Deviation (ubar)', 'FontSize', 12);
160    title(sprintf('Error Bars: X%cSEM', char(177)), 'FontSize', 10,...
```

```
            'FontWeight', 'bold');
161    print -dtiff Fig.B_ERROR-BARS-for-all-temperature-ranges.tif;


162    figure;
163    xii=[0 0 0 0 0 ];
164    ebGraph1=errorbar(xii,SEMi,'.r','LineWidth',1);
165    name2 = {' ';'T=0-15';'T=15-30';'T=30-40';'T=40-50';'T=50-60';...
       'T=0-60';' ';' '};
166    set(gca,'xticklabel',name2);
167    xlabel(sprintf('Temperature range (in %cC)', char(176)),...
           'FontSize', 12);
168    title(sprintf('Error Bars: SEM over zero value', char(177)),...
           'FontSize', 10, 'FontWeight', 'bold');
169    print -dtiff Fig.B2_ERROR-BARS-SEM_over-zero-value.tif;


170    P_dev1=1000000*abs(array1(:,1)-array1(:,3));
171    P_dev2=1000000*abs(array2(:,1)-array2(:,3));
172    P_dev3=1000000*abs(array3(:,1)-array3(:,3));
173    P_dev4=1000000*abs(array4(:,1)-array4(:,3));
174    P_dev5=1000000*abs(array5(:,1)-array5(:,3));
175    P_dev6=1000000*abs(Pressure(:,1)-Pressure(:,3));
176    boxes = cat(1, P_dev1, P_dev2, P_dev3, P_dev4, P_dev5, P_dev6);
177    set = [ones(size(array1(:,1))); 2*ones(size(array2(:,1)));...
           3*ones(size(array3(:,1))); 4*ones(size(array4(:,1)));...
           5*ones(size(array5(:,1))); 6*ones(size(Pressure(:,1)))];

178    for i=1:numel(boxes(:,1))
179      setValue=set(i,1);
180      switch setValue
181            case 1
182                    setStr(i,1)={'T=0-15'};
183            case 2
184                    setStr(i,1)={'T=15-30'};
185            case 3
186                    setStr(i,1)={'T=30-40'};
187            case 4
188                    setStr(i,1)={'T=40-50'};
```

```
189              case 5
190                      setStr(i,1)={'T=50-60'};
191              otherwise
192                      setStr(i,1)={'T=0-60'};
193      end
194  end


195  figure;
196  boxplot(boxes,setStr);
197  xlabel(sprintf('Temperature range (in %cC)', char(176)), 'FontSize',
     12);
198  ylabel('Pressure Deviation (ubar)', 'FontSize', 12);
199  title(sprintf('Box Plot: Spread and differences of samples at
     different temperature ranges'), 'FontSize', 10, 'FontWeight', 'bold');
200  print -dtiff Fig.C__BOX-PLOT-for-all-temperature-ranges.tif;


201  figure;
202  boxplot(boxes,setStr,'Notch','on');
203  xlabel(sprintf('Temperature range (in %cC)', char(176)), 'FontSize',
     12);
204  ylabel('Pressure Deviation (ubar)', 'FontSize', 12);
205  title(sprintf('Notched Box Plot: Sample Means and 95%c CI at different
     temperature ranges', char(37)), 'FontSize', 10, 'FontWeight', 'bold');
206  print -dtiff Fig.D__NOTCHED-BOX-PLOT-for-all-temperature-ranges.tif;


207 cd
```

The code lines 109–206 create five figures, for all of the six predefined temperature ranges (0–15°C, 15–30°C, 30–40°C, 40–50°C, 50–60°C, and 0–60°C), illustrating:

1. Distribution and sample mean of each temperature range (Fig.A__Distribution-for-all-temperature-ranges.tif);

2. Error bars of sample mean plus/minus SEM value (Fig.B__ERROR-BARS-for-all-temperature-ranges.tif);

3. Error bars of SEM over zero value (Fig.B2__ERROR-BARS-SEM_over-zero-value.tif);

4. Box plot diagrams of sample data for each particular temperature range (Fig.C__BOX-PLOT-for-all-temperature-ranges.tif);

5. Notched box plots of sample data for each particular temperature range (Fig.D__NOTCHED-BOX-PLOT-for-all-temperature-ranges.tif).

A brief description of these code lines is as follows. Code lines 109–139 plot the sample distribution along with mean value of sensors' deviation for each temperature range, using a *for-loop* statement (lines 110–140). All plots are depicted in the same figure. Within the loop, a *switch-case* statement makes a decision on the current array that will be plotted as well as the color of the current scheme according to the value of the loop counter; that is, red("r"), green("g"), blue("b"), cyan("c"), magenta("m"), and yellow("y") color. The—assumed to be normalized—distribution of the predefined color is plotted in lines 131–136, by plotting a histogram and fitting a normal density function (line 132) and, thereafter, deleting bars (line 133) of histogram. The sample mean value (of same color as in current distribution) is plotted in lines 137–139. Code lines 141–148 define x and y limits of the figure (lines 141–144) and append legend (line 145), y and x labels (lines 146, 147), and figure title (line 148). Code line 149 prints the figure in the default storage folder (in .tif format). The x and y limits are extended so that legend and distribution plots do not overlap each other, as the former incorporates several pieces of information, that is, the temperature range, sample mean, and p-p noise for each particular range.

Code lines 150–160 plot on the same figure the bar graphs of sample means (for all temperature ranges) along with error bars illustrating SEM, while line 161 stores the figure. Line 151 creates a cell of six rows holding the labels to be printed in the x axis. Line 152 creates a single-row array of six columns (denoted x) holding integer values from 1 to 6, which are associated with the number of temperature ranges and, hence, the position of each bar graph on x axis. Line 153 creates bar graphs of sample mean values, while line 154 replaces values 1–6 to the descriptive labels created before (i.e., line 151). Line 155 changes the color of bar graphs to black, while code lines 156, 157 append error bars of SEM around sample means. Lines 158–160 decide on the x, y, and title labels of the current figure. It is worth noting that *char(177)* corresponds to the *plus-minus* symbol (line 160) [1, 2]. Based on the same approach, code lines 162–169 depict error bars SEM around zero value (as determined by generated xii array of code line 163). The latter figure is addressed to graphically illustrate differences among SEM values of each temperature range.

Code lines 170–177 create two single-column arrays, named "boxes" and "set", which are later used to implement the box plot diagrams. The former array incorporates sensors' deviation (in ubar) for all temperature ranges (i.e., 0–15°C, 15–30°C, 30–40°C, 40–50°C, 50–60°C, 0–60°C and with this particular order). Sensors' deviation is determined by code lines 170–175 and overall values are concatenated in "box" (single-column) array in line 176. The "set" (single-column) array incorporates integer values 1–6, where each integer occupies a number of rows equivalent to the number of deviation values of each temperature range. Thereafter, the *switch-case* statement (lines 180–193) inside of the *for-loop* (lines 178–194) replaces integers to the descriptive string values of "setStr" cell. The box plot diagrams are generated in code lines 195–200, using information incorporated within "boxes" and "setStr", while code lines 201–206 visualize the same example using notched box plots (where the notch is used to show the 95% confidence interval for the median). Finally, code line 207 prints onto the command window the path of the current working folder (where the generated figure files are located).

A.2 MEASUREMENT ANALYSIS OF PRESSURE INFLUENCE (MATLAB CODE)

The following script code can be used as a reference guide for analyzing the influence on measurement accuracy in differential altimetry due to the sensed value of the existing pressure level (as explored in Chapter 3). The measurement files analyzed hereafter consist of 750 samples, where each level of pressure incorporates 125 measurements. Thereby, six sets of pressure level (i.e., from 0.960 bar to 0.985 bar with an increment of approximately 0.005 bar) are analyzed. To run the script code drag and drop the corresponding "Pressure.txt" into Matlab Workspace and then copy the code into the Matlab Command Window. The following code lines generate two figure files (that is, sensors' deviation as well as histogram of the latter signal) for each one of the six predefined pressure levels; plus two more figures for the overall data set (of 750 samples) and one figure depicting the data acquisition process. Thereby, 15 figures are created in the default working folder of Matlab, illustrating:

(a) Data acquisition process (Fig.0__Data-Acquisition-Process.tif);
(b) Deviation in sensors' output signal (Fig.x-1__Sensors-deviation-for-level-Px);
(c) Histogram of this deviation (Fig.x-2__Histogram-of-pressure-deviation.tif).

Number x in the filename of figures admits values in consideration of the consecutive number of the existing pressure level (that is, 1,2,3,…). A brief description of these code lines is as follows. The former two lines define the number of pressure levels employed in the measurement file (line 1) as well as the acquired samples at each particular level (line 2). Code lines 4–12 plot (lines 4–7) and store (line 12) the figure depicting the data acquisition process of barometer 1 (line 5) and barometer 2 (line 7) for the overall procedure. Labels of figure (i.e., x, y, legend, and title) are appended in the figure in code lines 8–11. Code lines 13–53 address a *for-loop* statement for plotting and storing figures of a) deviation in sensors' output (lines 27–38) and of b) corresponding histogram (lines 39–52), for each pressure level and for the overall procedure, as well.

Calculations of deviation are performed in code lines 14–26. The *if-else* statement (lines 14–20) extracts the samples of each pressure level (i.e., 1–125, 126–250, etc.) in code lines 17–19, while calculations for the overall set (i.e., 1–750) are decided during the last run of *for-loop* in code lines 14–16. At the end of the process seven arrays are generated (Pdev1-Pdev7), where each one of the six former arrays holds 125 samples of deviation in sensors' output signal for the pressure levels 0.960–0.985 bar, respectively, while the latter holds the 750 samples of deviation for all pressure levels. This particular operation is performed in code lines 21–25.

The seven means of deviation samples (individually calculated in line 24), are stored (in ubar) in the single-column array generated in code line 26. Plots of Pdev1-Pdev7 arrays are successively performed in code lines 27 and 28. Because the latter array consists of a different number of samples compared to arrays Pdev1–Pdev6, the range of x axis as well as title of this particular figure is accordingly modified in lines 34 and 35. Similarly, the title of the histogram figure is modified in code lines 46 and 47. Because the final calculations in the loop apply to the overall dataset, code line 53 returns (in Matlab Command Windows, since the ";" character is missing from this command) the maximum difference of deviation in sensors' output signal due to influence arising from the acquired pressure level.

```
1      Files=6
2      L=125;
3      histBins=10;

4      figure;
5      plot(Pressure(1:Files*L,1), 'r');
6      hold on;
7      plot(Pressure(1:Files*L,3), 'g');
8      xlabel('Pressure (ubar)', 'FontSize', 12);
9      ylabel('Samples', 'FontSize', 12);
10     legend('Sensor 1', 'Sensor 2', 'FontSize', 6);
11     title(sprintf('Data acquisition at different pressure levels...
          (T=%.1f %cC)' ,mean(Pressure(:,2)),char(176)),...
          'FontSize', 10, 'FontWeight', 'bold');
12     print -dtiff Fig.0__Data-Acquisition-Process.tif

13     for i=1:(Files+1)
14       if (i>Files)
15         idx1=1;
16         idx2=i*L-L;
17       else
18         idx1=i*L-L+1;
19         idx2=i*L;
20       end
21       P1=Pressure(idx1:idx2,1);
22       P2=Pressure(idx1:idx2,3);
23       Pdif=abs(P1-P2);
24       AvPdif=mean(Pdif);
```

```
25    eval(sprintf('Pdev%d = Pdif * 1000000', i));
26    AvPdev_ubar(i,1) = AvPdif*1000000;

27    figure;
28    plot( eval(sprintf('Pdev%d', i)) );
29    xlabel('Samples', 'FontSize', 12);
30    xlim([0 L]);
31    ylabel('Pressure (ubar)', 'FontSize', 12);
32    eval(sprintf('levelP%d = mean(P1)', i));
33    if (i>Files)
34     xlim([0 L*i-L]);
35     title(sprintf('Sensors%c deviation for level P:%.3f-%.3f bar...
          (T=%.1f %cC)', char(39),levelP1,levelP6,mean(Pressure(:,2)),...
          char(176) ), 'FontSize', 10, 'FontWeight', 'bold');
36    else
          title(sprintf('Sensors%c deviation for level P~%.3f bar (T=%.1f...
          %cC)', char(39),mean(P1(:,1)),mean(Pressure(:,2)), char(176) ),...
          'FontSize', 10, 'FontWeight', 'bold');
37    end
38    eval(['print -dtiff Fig.' num2str(i)...
          '-1__Sensors-deviation-for-level-Px' '.tif' ]);
39    figure;
40    histfit( eval(sprintf('Pdev%d', i)) ,histBins);
41    xlabel('Pressure Deviation (ubar)', 'FontSize', 12);
42    ylabel('Samples', 'FontSize', 12);
43    x=AvPdif*1000000;
44    s=std(Pdif) * 1000000;
45    ppNoise=6*s;
46    if (i>Files)
47     title(sprintf('Noise(p-p) = %.0fubar, Sample Mean = %.0fubar\n...
          level P:%.3f-%.3f bar (T=%.1f %cC)', ppNoise, AvPdev_ubar(i,1),...
          levelP1, levelP6, mean(Pressure(:,2)), char(176) ),...
          'FontSize', 10, 'FontWeight', 'bold');
48    else
49    title(sprintf('Noise(p-p) = %.0fubar, Sample Mean = %.0fubar\n...
          level P %c %.3f bar', ppNoise, AvPdev_ubar(i,1), char(126),...
          mean(P1(:,1)) ), 'FontSize', 10, 'FontWeight', 'bold');
50    end
51    eval(['print -dtiff Fig.' num2str(i)...
```

```
        '-2__Histogram-of-pressure-deviation-for-level-Px' '.tif' ]);

52   end

53   max(AvPdev_ubar)-min(AvPdev_ubar)
```

The code lines 54–125 create three figures for all of the seven predefined pressure ranges (i.e., 0.960 bar, 0.965 bar, 0.970 bar, 0.975 bar, 0.980 bar, 0.985 bar, as well the overall dataset of 0.960–0.985 bar), illustrating:

1. Distribution of deviation for each pressure level (Fig.A__Distribution-for-all-pressure-levels.tif);

2. Box plot diagrams of sample data for each particular level of pressure (Fig.B__BOX-PLOT-for-all-levels-of-pressure.tif);

3. Notched box plots of sample data for each particular level of pressure (Fig.C__NOTCHED-BOX-PLOT-for-all-levels-of-pressure.tif).

The process of generating these figures is similar to the one used in the previous script code example. A brief description of these code lines is as follows. Code lines 54–90 plot sample distribution sensors' deviation for each pressure level using a *for-loop* statement (lines 54–81). All plots are depicted in the same figure, while a *switch-case* statement (lines 56–71) inside of the loop makes a decision on the current array that will be plotted as well as the color of the current scheme according to the value of the loop counter. The color code is same as before, but also the key("k") color has been addressed for the additional (*7th*) distribution depicting the overall pressure levels.

Code lines 91–111 create two single-column arrays, named "boxes" and "set", which are later used to implement the box plot diagrams. As in the previous script code example, the former array incorporates sensors' deviation (in ubar) for all pressure levels, while the latter holds the integer values 1–7. Each integer occupies a number of rows equivalent to the number of deviation values of each level of pressure. Thereafter, the *switch-case* statement (lines 95–110) inside of the *for-loop* (lines 93–111) replaces integers to the descriptive string values included in the "setStr" cell. The box plot diagrams are generated in code lines 112–117, while code lines 118–124 visualize the same example using notched box plots. The generated figure files can be found in the path returned by code line 125.

```
54   figure;
55   for i=1:Files+1
56      switch i
57              case 1
```

```matlab
58                    color='r';
59            case 2
60                    color='g';
61            case 3
62                    color='b';
63            case 4
64                    color='c';
65            case 5
66                    color='m';
67            case 6
68                    color='y';
69            otherwise
70                    color='k';
71     end
72     hold on;
73     h=histfit( eval(sprintf('Pdev%d', i)) ,histBins);
74     delete(h(1));
75     hold on;
76     hLines = findobj('Type','Line');
77     set(hLines(1),'Color',color,'LineWidth',1);
78     xi(i)=mean(eval(sprintf('Pdev%d', i)));
79     pp(i)=6*std(eval(sprintf('Pdev%d', i)));
80     hold on;
81  end
82  xl=xlim;
83  xlim([ xl(1) (xl(2)*1.25) ]);
84  yl=ylim;
85  ylim([ yl(1) (yl(2)*1.001) ]);
86    legend( sprintf('P-level~%.3fbar\n x=%.0fubar,6*s=%.0fubar',...
      levelP1, xi(1), pp(1)),...
      sprintf('P-level~%.3fbar\n x=%.0fubar,6*s=%.0fubar',...
      levelP2, xi(2), pp(2)),...
      sprintf('P-level~%.3fbar\n x=%.0fubar,6*s=%.0fubar',...
      levelP3, xi(3), pp(3)),...
      sprintf('P-level~%.3fbar\n x=%.0fubar,6*s=%.0fubar',...
      levelP4, xi(4), pp(4)),...
      sprintf('P-level~%.3fbar\n x=%.0fubar,6*s=%.0fubar',...
      levelP5, xi(5), pp(5)),...
      sprintf('P-level~%.3fbar\n x=%.0fubar,6*s=%.0fubar',...
```

```matlab
           levelP6, xi(6), pp(6)),...
           sprintf('P-level:%.3f-%.3fbar\n x=%.0fubar,6*s=%.0fubar',...
           levelP1, levelP6, xi(7), pp(7)) );
87   ylabel('Samples', 'FontSize', 12);
88   xlabel('Pressure Deviation (ubar)', 'FontSize', 12);
89   title(sprintf('Distribution and sample mean at different pressure...
      levels (T=%.1f %cC)' ,mean(Pressure(:,2)),char(176)),...
      'FontSize', 10, 'FontWeight', 'bold');
90   print -dtiff Fig.A__Distribution-for-all-pressure-levels.tif;
91   boxes = cat(1, Pdev1, Pdev2, Pdev3, Pdev4, Pdev5, Pdev6, Pdev7);
92   set = [ones(size(Pdev1(:,1))); 2*ones(size(Pdev2(:,1)));...
      3*ones(size(Pdev3(:,1))); 4*ones(size(Pdev4(:,1)));...
      5*ones(size(Pdev5(:,1))); 6*ones(size(Pdev6(:,1)));...
      7*ones(size(Pdev7(:,1)))];
93   for i=1:numel(boxes(:,1))
94    setValue=set(i,1);
95    switch setValue
96     case 1
97     setStr(i,1)={sprintf('%.3f', eval(sprintf('levelP%d', setValue)) )};
98     case 2
99     setStr(i,1)={sprintf('%.3f', eval(sprintf('levelP%d', setValue)) )};
100    case 3
101    setStr(i,1)={sprintf('%.3f', eval(sprintf('levelP%d', setValue)) )};
102    case 4
103    setStr(i,1)={sprintf('%.3f', eval(sprintf('levelP%d', setValue)) )};
104    case 5
105    setStr(i,1)={sprintf('%.3f', eval(sprintf('levelP%d', setValue)) )};
106    case 6
107    setStr(i,1)={sprintf('%.3f', eval(sprintf('levelP%d', setValue)) )};
108    otherwise
109    setStr(i,1)={sprintf('overall')};
110   end
111  end


112  figure;
113  boxplot(boxes,setStr);
114  xlabel(sprintf('Pressure levels inside of the airtight enclosure',...
      char(176)), 'FontSize', 12);
```

```
115  ylabel('Pressure Deviation (ubar)', 'FontSize', 12);
116  title(sprintf('Box Plot: Spread and differences of samples at...
        different pressure levels'), 'FontSize', 10, 'FontWeight', 'bold');
117  print -dtiff Fig.B__BOX-PLOT-for-all-levels-of-pressure.tif


118  figure;
119  boxplot(boxes,setStr,'Notch','on');
120  xlabel(sprintf('Pressure levels inside of the airtight enclosure',...
        char(176)), 'FontSize', 12);
121  ylabel('Pressure Deviation (ubar)', 'FontSize', 12);
122  title(sprintf('Notched Box Plot: Sample Means and 95%c CI at...
123   different pressure levels', char(37)), 'FontSize', 10,...
        'FontWeight', 'bold');
124  print -dtiff Fig.C__NOTCHED-BOX-PLOT-for-all-levels-of-pressure.tif

125  cd
```

A.3   STATISTICAL ANALYSIS OF HEIGHT MEASUREMENTS (MATLAB CODE)

The following script code can be used as a reference guide for analyzing absolute height measurements in differential and single-ended altimetry. There are two measurement files analyzed hereafter, consisting of 1,200 measurement each. The file named "PressureSame.txt" embeds measurements arisen from the two barometers, when both devices were placed at the same height (i.e., reference position). The file named "PressureDifferent.txt" is in line with measurements obtained from the two barometers when the latter were placed at different elevations. The two files include data from 10 identical measurement procedures and, hence, each set consists of 120 samples. To run the script code drag and drop both files into Matlab Workspace and then copy the code into the Matlab Command Window. The following code lines generate two figure files depicting the overall data acquisition process at reference position and height difference, respectively. The name of each file is as follows:

(a) Fig.1Pressure_data_at_referece_position.tif;

(b) Fig.2Pressure_data_at_height_difference.tif.

A brief description of these code lines 1–79 is as follows. The former two lines define the number of sets as well as samples of each set, respectively, while the third line defines the population mean value of absolute height evaluated by the altimetry methods. Lines 4–6 specify the constants used by the hypsometric equation. Plots of the two figures are addressed within a *for-loop* (lines 8–28). The latter statement runs twice (line 8) and the first time it plots measurements at reference position (as determined by the *if-else* statement of code lines 9–13). Accordingly, the title of the figure is determined by the corresponding *if-else* statement of code lines 21–27.

The *for-loop* statement of code lines 29–79 calculates the absolute height difference for all of the 10 sets of measurement, in differential and single-ended altimetry using both the *international barometric formula* (IBF) and *hypsometric equation* (HE). In detail, code lines 30–37 calculate the deviation in sensors' output signal (line 36) and its average value (line 37). Code lines 38–42 calculate the scale height (H) which is later used by the hypsometric equation. The value of temperature is determined by averaging measurements obtained from both barometers at height difference (lines 38–41), thereby assuming an isothermal atmosphere in between sensor devices.

Absolute height determination in differential altimetry is performed in code lines 43–62. Corrections to the acquired pressure samples, because of the unwanted sensors' deviation, are applied to the base barometer. If the acquired pressure of the rover barometer at reference position is greater than the one of the base altimeter, corrections to the latter device are in line with a boost relative to the assessed deviation (lines 44–46). Otherwise, the average value of deviation is subtracted from each single pressure sample acquired by the base altimeter at reference position (lines 49–51). Height determination using IBF is, respectively, performed in code lines 47 and 52, while the corresponding calculation using HE is fetched in lines 48 and 53. Additional calculations are performed in code lines 55–62. Code lines 55 and 56 assign to single-column arrays hIBFn and HEn the height determination based on IBF and HE calculations, respectively, where, n is an integer identical to the loop counter value being associated with the current set of measurements. In terms of the IBF calculations, sample mean and standard deviation of the above array as well as the percent error in height determination (relative to the population mean) is fetched in code lines 57, 58, and 61, respectively. The corresponding HF calculations are performed in code lines 59, 60, and 62. Similarly, the absolute height determination in single-ended altimetry is performed in code lines 64–78.

```
1     Files=10;
2     L=120;
3     u=0.885;
4     k=1.38*power(10,-23);
5     g=9.81;
6     m=(0.22*(2*2.67*power(10,-26))) + (0.78*(2*2.3*power(10,-26)));
7     histBins=10;


8     for i=1:2

9       if (i==1)
10        file=PressureSame;
```

```
11    else
12     file=PressureDifferent;
13    end
14    figure;
15    plot(file(1:Files*L,1), 'r');
16    hold on;
17    plot(file(1:Files*L,3), 'g');
18    ylabel('Pressure (bar)', 'FontSize', 12);
19    xlabel('Samples', 'FontSize', 12);
20    legend('BASE', 'ROVER', 'FontSize', 6);
21    if (i==1)
22    title(sprintf('Data acquisition at reference position...
         (T=%.1f %cC)' , (( mean(file(:,2))+ mean(file(:,4)) )/2),...
         char(176)), 'FontSize', 10, 'FontWeight', 'bold');
23      eval(['print -dtiff Fig.' num2str(i)...
           'Pressure_data_at_referece_position.tif' ]);
24    else
25      title(sprintf('Data acquisition at height difference...
           (T=%.1f %cC)', (( mean(file(:,2))+ mean(file(:,4)) )/2) ,...
           char(176)), 'FontSize', 10, 'FontWeight', 'bold');
26      eval(['print -dtiff Fig.' num2str(i)...
           'Pressure_data_at_height_difference.tif' ]);
27    end

28    end


29    for i=1:Files
30    idx1=1;
31    idx2=i*L;
32    BASE1=PressureSame(idx1:idx2,1);
33    BASE2=PressureDifferent(idx1:idx2,1);
34    ROVER1=PressureSame(idx1:idx2,3);
35    ROVER2=PressureDifferent(idx1:idx2,3);
36    Pdif=abs(BASE1-ROVER1);
37    PdifAv=mean(Pdif);
38    T1 = mean(PressureDifferent(idx1:idx2,2)) + 273.15;
39    T2 = mean(PressureDifferent(idx1:idx2,4)) + 273.15;
40    T=(T1+T2)/2;
```

```matlab
41    ToC=T-273.15;
42    H=(k*T)/(m*g);

43    % Differential Altimetry
44    if ( mean(BASE1) < mean(ROVER1) )
45     BASE1=BASE1+PdifAv;
46     BASE2=BASE2+PdifAv;
47     hIBF=4935.125 * ( power((ROVER2*100000),0.1903)...
        - power((BASE2*100000),0.1903) );
48     hHE=H*(log(ROVER2*100000)-log(BASE2*100000));
49    else
50     BASE1=BASE1-PdifAv;
51     BASE2=BASE2-PdifAv;
52     hIBF=4935.125 * ( power((BASE2*100000),0.1903)...
        - power((ROVER2*100000),0.1903) );
53     hHE=H*(log(BASE2*100000)-log(ROVER2*100000));
54    end
55    eval(sprintf('hIBF%d = hIBF', i));
56    eval(sprintf('hHE%d = hHE', i));
57    xi_hIBF(i)=mean(eval(sprintf('hIBF%d', i)));
58    si_hIBF(i)=std(eval(sprintf('hIBF%d', i)));
59    xi_hHE(i)=mean(eval(sprintf('hHE%d', i)));
60    si_hHE(i)=std(eval(sprintf('hHE%d', i)));
61    error_IBF(i)=( abs(xi_hIBF(i)-u) / u ) * 100;
62    error_HE(i)=( abs(xi_hHE(i)-u) / u ) * 100;

63    % Single-ended Altimetry
64    if ( mean(ROVER2) < mean(ROVER1) )
65     hIBFs=4935.125 * ( power((ROVER1*100000),0.1903)...
        - power((ROVER2*100000),0.1903) );
66     hHEs=H*(log(ROVER1*100000)-log(ROVER2*100000));
67    else
68     hIBFs=4935.125 * ( power((ROVER2*100000),0.1903)...
        - power((ROVER1*100000),0.1903) );
69     hHEs=H*(log(ROVER2*100000)-log(ROVER1*100000));
70    end
71    eval(sprintf('hIBFs%d = hIBFs', i));
72    eval(sprintf('hHEs%d = hHEs', i));
73    xi_hIBFs(i)=mean(eval(sprintf('hIBFs%d', i)));
```

```
74   si_hIBFs(i)=std(eval(sprintf('hIBFs%d', i)));
75   xi_hHEs(i)=mean(eval(sprintf('hHEs%d', i)));
76   si_hHEs(i)=std(eval(sprintf('hHEs%d', i)));
77   error_IBFs(i)=( abs(xi_hIBFs(i)-u) / u ) * 100;
78   error_HEs(i)=( abs(xi_hHEs(i)-u) / u ) * 100;

79   end
```

The code lines 80–132 create two figures incorporating box plot diagrams for all of the ten sets of height determination, with the latter being calculated with (a) IBF and (b) HE formula, respectively. In addition, this part of the code generates three measurement files as well. The generated file are as follows:

1. Box plots of height graph with IBF (Fig.1BOX-PLOT-for-DBA.tif);

2. Box plots of height graph with HE (Fig.2BOX-PLOT-for-DBA.tif);

3. Measurement results returned from differential altimetry (resuts_DBA.txt);

4. Measurement results returned from single-ended altimetry (resuts_SBA.txt);

5. Summary of height determination arisen from both altimetry methods (resuts_SBA.txt).

The former two files consist of eleven rows and six columns. The first ten rows incorporate data for each particular set of height determination, while the final row integrates mean values of each column. The first two columns are in agreement with the sample mean and sample standard deviation of absolute height, with the latter being determined with IBF. The next two columns incorporate the same information for absolute height, with the latter being determined with HE. The final two columns provide the percent error in height determination, when the experimental value of height is obtained from IBF and HE, respectively. The last file consists of four columns and ten rows (for each one of the ten sets of height). The first two columns employ height results arisen from IBF in differential and single-ended altimetry, respectively. The latter two columns employ height results arisen from HE in differential and single-ended altimetry, respectively.

A brief description of these code lines is as follows. Code lines 80–98 generate box plot diagrams of the ten measurement sets of absolute height in differential barometric altimetry, within a *for-loop* running twice. The first time the loop runs, it generates box plots of height being determined by IBF. The second time height determination is in agreement with the HE. This action is decided by the *if-else* statement of code lines 81–85. Because hIBFn and hHEn are identical arrays, size of each set is determined only one time, in consideration of the former array size (that is, code line 86). Thereby, in the first run of the loop the single-column arrays named hIBFn are concatenated in code line 82. In the second run of the loop, concatenation is performed to hHEn arrays (line 84). Plotting process is similar to the one addressed by the previous examples. In addition, code lines 93 and 94 append a horizontal dashed line of green color parallel to the x axis, illustrating the population mean value (i.e., 0.885 ubar for this particular example). To generate notched box plots instead of the regular box plots, we need to uncomment line 89 and comment the precede code line (i.e., 88).

```
80    for i=1:2
81     if (i==1)
82      boxes = cat(1, hIBF1,hIBF2,hIBF3,hIBF4,hIBF5,...
         hIBF6,hIBF7,hIBF8,hIBF9,hIBF10);
83     else
84      boxes = cat(1, hHE1,hHE2,hHE3,hHE4,hHE5,...
         hHE6,hHE7,hHE8,hHE9,hHE10);
85     end
86     set = [ones(size(hIBF1));2*ones(size(hIBF2));...
         3*ones(size(hIBF3));4*ones(size(hIBF4));5*ones(size(hIBF5));...
         6*ones(size(hIBF6));7*ones(size(hIBF7));8*ones(size(hIBF8));...
         9*ones(size(hIBF9));10*ones(size(hIBF10))];
87     figure;
88     boxplot(boxes,set);
89     %boxplot(boxes,set,'Notch','on');
90     xlabel(sprintf('Measurement set', char(176)), 'FontSize', 12);
91     ylabel('Absolute height (m)', 'FontSize', 12);
92     title(sprintf('Boxplots for %.3fm height estimation',u),...
         'FontSize', 10, 'FontWeight', 'bold');
93     hold on;
94     plot(xlim, [u,u],'--g','LineWidth',1);
95     legend(sprintf('population mean = %.3fm',u));
96     %print -dtiff Fig.A__BOX-PLOT-for-DBA-using-IBF.tif
97     eval(['print -dtiff Fig.' num2str(i)...
         'BOX-PLOT-for-DBA.tif' ]);
98    end

99    % Differential Altimetry results
100   x_hIBF=mean(xi_hIBF)
101   s_hIBF=mean(si_hIBF)
102   x_hHE=mean(xi_hHE)
103   s_hHE=mean(si_hHE)

104   fileID = fopen('resuts_DBA.txt','w');
```

```
105   for i=1:numel(xi_hIBF)
106    fprintf(fileID,'%.2f\t %.2f\t %.2f\t %.2f\t %.1f\t %.1f\r\n',...
         xi_hIBF(i),si_hIBF(i),xi_hHE(i),si_hHE(i),error_IBF(i),error_HE(i));
107   end
108   fprintf(fileID,'\r\n\r\n\r\n%.2f\t %.2f\t %.2f\t %.2f\t %.2f\t %.2f\t...
        %.1f\t %.1f\r\n',...
        x_hIBF,s_hIBF,x_hHE,s_hHE,mean(error_IBF),mean(error_HE));
109   fclose(fileID);


110   %One-sample t-test
111   [h,p,ci,stats] = ttest(xi_hIBF,u)
112   [h,p,ci,stats] = ttest(xi_hHE,u)



113   % Single-ended Altimetry results
114   x_hIBFs=mean(xi_hIBFs)
115   s_hIBFs=mean(si_hIBFs)
116   x_hHEs=mean(xi_hHEs)
117   s_hHEs=mean(si_hHEs)


118   fileID = fopen('resuts_SBA.txt','w');
119   for i=1:numel(xi_hIBFs)
120    fprintf(fileID,'%.2f\t %.2f\t %.2f\t %.2f\t %.1f\t %.1f\r\n',...
         xi_hIBFs(i),si_hIBFs(i),xi_hHEs(i),si_hHEs(i),...
         error_IBFs(i),error_HEs(i));
121   end
122   fprintf(fileID,'\r\n\r\n\r\n%.2f\t %.2f\t %.2f\t %.2f\t...
        %.1f\t %.1f\r\n',...
        x_hIBFs,s_hIBFs,x_hHEs,s_hHEs,...
        mean(error_IBFs),mean(error_HEs));
123   fclose(fileID);
124   %One-sample t-test
125   [h,p,ci,stats] = ttest(xi_hIBFs,u)
126   [h,p,ci,stats] = ttest(xi_hHEs,u)


127   % Overall measurements
128   fileID = fopen('results.txt','w');
129   for i=1:numel(xi_hIBFs)
```

```
        fprintf(fileID,'%.2f\t %.2f\t %.2f\t %.2f\r\n',...
        xi_hIBF(i),xi_hIBFs(i),xi_hHE(i),xi_hHEs(i));
130   end
131   fclose(fileID);

132   cd
```

Some additional calculation of average values in differential barometric altimetry are performed in code lines 100–103 and, in particular, the mean value of height determined by IBF (line 100), the mean value of standard deviation of the latter signal (line 101), as well as the corresponding values of height determined by HE (lines 102 and 103, respectively). The file resuts_DBA.txt in the scheme described before is created in code lines 104–109 and stored to the default working folder of Matlab software. One-sample t-test on the differential altimetry method using IBF and HE calculations are, respectively, performed in code lines 111 and 112. Results returned from these two tests are printed on the Matlab Command Window console. The latter actions are repeated for the single-barometric altimetry method in code lines 113–126, and the summary of height determination arisen from both methods of altimetry are stored to resuts_SBA.txt file in code lines 128–131.

REFERENCES

[1] List of Unicode characters, https://en.wikipedia.org/wiki/List_of_Unicode_characters. [Accessed: Mar-2017].

[2] Latin-1 Supplement (Unicode block), https://en.wikipedia.org/wiki/Latin-1_Supplement_(Unicode_block). [Accessed: Mar-2017].

Abbreviations

ASCII	American Standard Code for Information Interchange
BFSL	Best-Fit Straight-Line
BPS	Barometric Pressure Sensor
CAD	Computer Aided Design
CI	Confidence Interval
DUT	Device Under Test
FFD	Full-Function Device
FS	Full Scale (span)
GPS	Global Positioning System
GUI	Graphical User Interface
HE	Hypsometric Equation
IBF	International Barometric Formula
IC	Integrated Circuit
I2C	Inter-Integrated Circuit
IMU	Inertial Measurement Unit
IoT	Internet of Things
IPIN	Indoor Positioning and Indoor Navigation
IQR	Inter-quartile Range
ISA	International Standard Atmosphere
LED	Light Emitting Diode
MEMS	Micro-Electro-Mechanical-Systems
PAN	Personal Areas Network
PCB	Printed Circuit Board
Q	Quartile
QNH	Query Nautical Heigh (referred to local seal level pressure)
RFD	Reduced-Function Device
SEM	Standard Error of the Mean
SiP	System in Package
SPI	Serial Peripheral Interface
TC	Temperature Coefficient
TEB	Total Error Band
WLP	Wafer-Level Packaging
WSN	Wireless Sensor Network

Author's Biography

DIMOSTHENIS E. BOLANAKIS

Dimosthenis E. Bolanakis was born in Crete, Greece, in 1978. He obtained a B.Sc. degree in Electronic Engineering from the Dept. of Electronics, Thessalonikis Educational Institute of Technology, Greece, an M.Sc. degree in Modern Electronic Technologies from the Dept. of Physics, University of Ioannina, Greece, and a Ph.D. degree from the Department of Primary Education, University of Ioannina, Greece (PhD Thesis: "Research and assessment of remote experiments in physics education using wireless networks").

D. E. Bolanakis has (co)authored more than 30 papers (mainly on Research in Engineering Education) and he is one of the authors of *Microcomputer Architecture: Low-level Programming Methods and Applications of the M68HC908GP32*. He has refereed articles for the *IEEE Transactions on Education* (IEEE), *Computer Applications in Engineering Education* (WILEY), *International Journal of Engineering Education* (TEMPUS), while he has participated in research projects for a) Designing and Implementing (FPGA-based) Digital Mammography Systems, b) Reinforcing Informatics' Education, and c) Broadening Higher Education. He has been occupied as an electronic engineer in the industry from 2010–2014, while during the period 2012–2014 he joined the European System Sensors S.A. (http://www.esenssys.com/) corporation that specialized in the design of MEMS sensors. He has worked as a laboratory associate at the Dept. of Informatics and Telecommunications, Epirus Educational Institute of Technology, Greece, for the teaching of Computer Architecture course (years: 2003–2009 with 2,334 teaching hours), as well as a teaching assistant at the Dept. of Physics, University of Ioannina, Greece, for the teaching of Microcontrollers – Microprocessors course (years: 2003–2004). His research interests include a) MEMS Sensors System-level Design and Measurement Analysis, b) μC-based & FPGA-based Digital Hardware Design, and c) Research in Engineering Education. Currently he holds a Special Lab & Teaching Personnel position at Hellenic Air Force Academy (Informatics & Computers section).

Printed in the United States
by Baker & Taylor Publisher Services